M. Sandra Wood, MLS, MBA
Janet M. Coggan, MEd, MSLS
Editors

Women's Health
on the Internet

Women's Health on the Internet has been co-published simultaneously as *Health Care on the Internet,* Volume 4, Numbers 2/3 2000.

Pre-publication
REVIEWS,
COMMENTARIES,
EVALUATIONS . . .

"**T**his volume provides thoughtful analysis of the issues surrounding the evaluation of Web resources for women and a discussion of the cultural concepts surrounding 'women's health' in general and how these ideas are played out on the Web. It will assist librarians who want to develop their own user guides and pathfinders, and will be extremely helpful for creating presentations on women's health and resources."

Margaret Bandy, MALS, AHIP
Manager, Exempla Healthcare Libraries

"This collection of articles provides an authoritative guide to the wealth of Internet information available on topics associated with women's health. Prepared by knowledgeable health sciences librarians, this work describes issues related to women's health, and provides timesaving lists of excellent sites and guidance on how to evaluate Internet sites and search engines.

This book will be useful to all health consumers and to those collections supporting consumer health information."

Karen Graves, MLS, AHIP
Head, Access Services
Library of the Health Sciences
University of Illinois at Chicago

"This valuable resource selects the best from the flood of Web sites providing health information for women. Covering broad as well as specialized sites and resources for the novice as well as the seasoned Web surfer, this book should be in every public and consumer health library. It will serve as a bible to hospital librarians conducting women's health information seminars for their institutions' clients. Physicians and nurses can confidently recommend this book to their patients. All in all, a thorough, well-written book."

Lillian R. Brazin, MS, AHIP
Director, Library Services
Albert Einstein Healthcare Network
Philadelphia, Pennsylvania

"This is a much-needed volume in the area of women's health. As more and more consumers turn to the Internet for answers, this book will prove invaluable. There is no such compilation currently available elsewhere. The chapters are uniformly designed and written in language consumers will understand.

Although geared to the consumer, this work will also be an excellent resource for medical librarians. The resources are well-documented, referenced, and selected with care.

All in all, this book will make an excellent addition to a reference collection in either a medical or public library as well as a handbook for the consumer."

Frances A. Brahmi, MA, MLS
Curriculum and Education Director
Ruth Lilly Medical Library
Indiana University School of Medicine
Indianapolis, Indiana

Women's Health
on the Internet

Women's Health on the Internet has been co-published simultaneously as *Health Care on the Internet,* Volume 4, Numbers 2/3 2000.

The *Health Care on the Internet* Monographic "Separates"

Below is a list of "separates," which in serials librarianship means a special issue simultaneously published as a special journal issue or double-issue *and* as a "separate" hardbound monograph. (This is a format which we also call a "DocuSerial.")

"Separates" are published because specialized libraries or professionals may wish to purchase a specific thematic issue by itself in a format which can be separately cataloged and shelved, as opposed to purchasing the journal on an on-going basis. Faculty members may also more easily consider a "separate" for classroom adoption.

"Separates" are carefully classified separately with the major book jobbers so that the journal tie-in can be noted on new book order slips to avoid duplicate purchasing.

You may wish to visit Haworth's website at . . .

http://www.HaworthPress.com

. . . to search our online catalog for complete tables of contents of these separates and related publications.

You may also call 1-800-HAWORTH (outside US/Canada: 607-722-5857), or Fax 1-800-895-0582 (outside US/Canada: 607-771-0012), or e-mail at:

getinfo@haworthpressinc.com

Women's Health on the Internet, edited by M. Sandra Wood, MLS, MBA, and Janet M. Coggan, MEd, MSLS (Vol. 4, No. 2/3, 2000). *A guide to Web sites representing women's health organizations, sites where you can find current women's health news, and much more.*

HIV/AIDS Internet Information Sources and Resources, edited by Jeffrey T. Huber, PhD (Vol. 2, No. 2/3, 1998). *"Will demystify and detangle the World Wide Web for the AIDS researcher, the primary care provider, and the patient anxious for first-hand information in various levels of detail." (Kiyoshi Kuromiya, Director, Critical Path AIDS Project, Philadelphia, Pennsylvania)*

Cancer Resources on the Internet, edited by M. Sandra Wood, MLS, MBA, and Eric P. Delozier, MLS (Vol. 1, No. 3, 1997). *"A refreshing and welcome emphasis on what's on the Internet for medicine consumers together with a wealth of useful resources and techniques." (Alan M. Rees, Professor Emeritus, Case Western Reserve University, Editor, Consumer Health and Nutrition Index, Author, Consumer Health Information Source Book, 4th Edition)*

Women's Health on the Internet

M. Sandra Wood, MLS, MBA
Janet M. Coggan, MEd, MSLS
Editors

Women's Health on the Internet has been co-published simultaneously as *Health Care on the Internet,* Volume 4, Numbers 2/3 2000.

The Haworth Press, Inc.
New York • London • Oxford

Women's Health on the Internet has been co-published simultaneously as *Health Care on the Internet,* Volume 4, Numbers 2/3 2000.

The development, preparation, and publication of this work has been undertaken with great care. However, the publisher, employees, editors, and agents of The Haworth Press and all imprints of The Haworth Press, Inc., including The Haworth Medical Press® and The Pharmaceutical Products Press®, are not responsible for any errors contained herein or for consequences that may ensue from use of materials or information contained in this work. Opinions expressed by the author(s) are not necessarily those of The Haworth Press, Inc.

Cover design by Thomas J. Mayshock Jr.

Library of Congress Cataloging-in-Publication Data

Women's health on the internet / M. Sandra Wood, Janet M. Coggan, editors.
     p. cm.
     "Women's health on the internet has been co-published simultaneously as Health care on the internet, volume 4, Numbers 2/3 2000."
     Includes bibliographical references and index.
     ISBN 0-7890-1300-2 (alk. paper)–ISBN 0-7890-1301-0 (alk. paper)
     1. Women–Health and hygiene–Computer network resources. 2. Internet. 3. Medical informatics.
I. Wood, M. Sandra. II. Coggan, Janet M.
RA778 .W756 2000
025.06'6130424–dc21   00-058213

INDEXING & ABSTRACTING

Contributions to this publication are selectively indexed or abstracted in print, electronic, online, or CD-ROM version(s) of the reference tools and information services listed below. This list is current as of the copyright date of this publication. See the end of this section for additional notes.

- *Abstracts in Social Gerontology: Current Literature on Aging*

- *Adis International Ltd.*

- *AgeLine Database*

- *Applied Social Sciences Index & Abstracts (ASSIA) (Online: ASSI via Data-Star) (CDRom: ASSIA Plus)*

- *Brown University Digest of Addiction Theory and Application, The (DATA Newsletter)*

- *BUBL Information Service, an Internet-based Information Service for the UK higher education community <URL:http://bubl.ac.uk/>*

- *Cambridge Scientific Abstracts*

- *CINAHL (Cumulative Index to Nursing & Allied Health Literature), in print, also on CD-ROM from CD Plus, EBSCO, and SilverPlatter, and online from CDP Online (formerly BRS), Data-Star, and PaperChase. (Support materials include Subject Heading List, Database Search Guide, and instructional video.)*

- *CNPIEC Reference Guide: Chinese National Directory of Foreign Periodicals*

- *Combined Health Information Database (CHID)*

- *Computing Reviews*

- *Current Awareness Abstracts of Library & Information Management Literature, ASLIB (UK)*

- *European Association for Health Information & Libraries: selected abstracts in newsletter "Publications" section*

(continued)

- *FINDEX <www.publist.com>*
- *Health Care Literature Information Network/HECLINET*
- *Health Service Abstracts (HSA)*
- *Healthcare Leadership Review*
- *Index to Periodical Articles Related to Law*
- *Information Science Abstracts*
- *INSPEC*
- *Internet & Personal Computing Abstracts <www.infotoday.com/mca/default.htm>*
- *Journal of the American Dietetic Association (Abstract Section)*
- *Leeds Medical Information*
- *Library Association Health Libraries Group Newsletter*
- *Library & Information Science Abstracts (LISA)*
- *Library and Information Science Annual (LISCA) <www.lu.com/arba>*
- *Medicinal & Aromatic Plants Abstracts (MAPA)*
- *OT BibSys*
- *PASCAL, c/o Insititute de L'Information Scientifique et Technique <http://www.inist.fr>*
- *Patient Care Management Abstracts*
- *Pharmacy Business*
- *Referativnyi Zhurnal (Abstracts Journal of the All-Russian Institute of Scientific and Technical Information–in Russian)*
- *Social Work Abstracts*

(continued)

Special Bibliographic Notes related to special journal issues (separates) and indexing/abstracting:

- indexing/abstracting services in this list will also cover material in any "separate" that is co-published simultaneously with Haworth's special thematic journal issue or DocuSerial. Indexing/abstracting usually covers material at the article/chapter level.
- monographic co-editions are intended for either non-subscribers or libraries which intend to purchase a second copy for their circulating collections.
- monographic co-editions are reported to all jobbers/wholesalers/approval plans. The source journal is listed as the "series" to assist the prevention of duplicate purchasing in the same manner utilized for books-in-series.
- to facilitate user/access services all indexing/abstracting services are encouraged to utilize the co-indexing entry note indicated at the bottom of the first page of each article/chapter/contribution.
- this is intended to assist a library user of any reference tool (whether print, electronic, online, or CD-ROM) to locate the monographic version if the library has purchased this version but not a subscription to the source journal.
- individual articles/chapters in any Haworth publication are also available through the Haworth Document Delivery Service (HDDS).

Women's Health on the Internet

CONTENTS

ABOUT THE EDITORS

M. Sandra Wood, MLS, MBA, is Librarian, Reference and Database Services, at The Milton S. Hershey Medical Center of the Pennsylvania State University at Hershey. She has over 29 years of experience as a medical reference librarian. Ms. Wood has published widely in the field of medical reference and is Editor of the journals *Medical Reference Services Quarterly* and *Health Care on the Internet.* She is active in the Medical Library Association and Special Libraries Association, most recently serving on MLA's Board of Directors as Treasurer.

Janet M. Coggan, MEd, MSLS, is a consultant located in Gainesville, Florida. She has been involved in medical librarianship for over a decade. Widely known in academic health and medical publishing for her book reviews, she has been Book Review Editor for the journals *Medical Reference Services Quarterly* and the *Bulletin of the Medical Library Association* and is Associate Editor of the journal *Health Care on the Internet.*

Women's Health on the Web:
An Overview

Dolores Zegar Judkins

ABSTRACT. With the plethora of women's health information on the Web, it is very easy to find oneself wallowing in thousands of Web sites when searching for women's health information. This is an attempt at gathering together some of the best of the Web in general women's health. Sites included are: sites with original material, "webliographies," women's health organizations, current news, and general sites for specific groups of women. *[Article copies available for a fee from The Haworth Document Delivery Service: 1-800-342-9678. E-mail address: <getinfo@haworthpressinc.com> Website: <http://www.HaworthPress.com>]*

KEYWORDS. Women's health, gynecology, consumer health, Internet, adolescent health

INTRODUCTION

Surveys have repeatedly shown that one of the major reasons women use the Internet is to find health information. Searches on the Internet retrieve thousands to hundreds of thousands of women's health sites, with informa-

Dolores Zegar Judkins (judkinsd@ohsu.edu) received her MLS from the University of Oregon in 1973. She is currently Librarian and Web Manager for the Oregon Health Sciences University Center for Women's Health, and Reference Librarian, Health Sciences Library, Oregon Health Sciences University, P.O. Box 573, Portland, OR 97207-0573. She was formerly Coordinator of Consumer Health Resources at OHSU.

[Haworth co-indexing entry note]: "Women's Health on the Web: An Overview." Judkins, Dolores Zegar. Co-published simultaneously in *Health Care on the Internet* (The Haworth Press, Inc.) Vol. 4, No. 2/3, 2000, pp. 1-18; and: *Women's Health on the Internet* (ed: M. Sandra Wood, and Janet M. Coggan) The Haworth Press, Inc., 2000, pp. 1-18. Single or multiple copies of this article are available for a fee from The Haworth Document Delivery Service [1-800-342-9678, 9:00 a.m. - 5:00 p.m. (EST). E-mail address: getinfo@haworthpressinc.com].

tion ranging from very specific topics to comprehensive sites. Sites are produced by individuals, government agencies, hospitals, universities, and everyone in between. This article is an attempt to gather sites that present a
broad spectrum of women's health information, in easy-to-use formats, for
consumers, health practitioners, and researchers. There is no attempt at being
comprehensive, as well as no attempt to list Web sites in specific subject
categories.

The sites are organized in broad categories. The first group is a list of three
major sites that have links to information on virtually every women's health
topic. Following are Web sites that have full-text information, and then "webliographies," or sites with links to many other sites. Next is a section on
search engine directories with women's health sections. Sections on government resources, women's health organizations, and news sites complete the
general women's health information. The final sections are for specific categories of women: young women, senior women, and a special list including
ethnic, lesbian, disabled, and rural women.

WOMEN'S HEALTH MEGASITES

These sites include almost every aspect of women's health issues in a
variety of formats. They all have full-text articles, include links to other sites,
and have lists of organizations. These are places to turn to first when looking
for general or specific information on women's health.

MEDLINE*plus*: Women's Health
<http://www.nlm.nih.gov/medlineplus/womenshealth.html>

From the National Library of Medicine, this site offers a plethora of links
to government, educational, and commercial sites on a large variety of
women's health topics. Almost all subjects have Spanish language information included. Links to pre-done PubMed searches are included. MEDLINE-
plus is a definite first-stop source.

healthfinder: women
<http://www.healthfinder.gov/justforyou/women/default.htm>

Jointly sponsored by healthfinder, from the Department of Health and
Human Services and the National Women's Health Information Center, this
site includes links to full-text information on a variety of women's health
issues, news, organizations, how to make smart health choices, and resources
for specific groups such as working mothers, lesbian parents, and retirees.
The search engine for the entire healthfinder database is also available.

NOAH: Ask NOAH About Women's Health
<http://www.noah.cuny.edu/healthyliving/womenshealth.html>

A project by The City University of New York, The Metropolitan New York Library Council, The New York Academy of Medicine, and The New York Public Library, this site is of particular interest because many of the articles are in both English and Spanish. Included are general articles about women's health and anatomy, as well as articles on specific topics such as menopause, osteoporosis, and cancer.

FULL-TEXT INFORMATION ON WOMEN'S HEALTH

The following sites are from a variety of sources. Some are from medical institutions, some from commercial Web sites. All of them have original information in full text on a broad spectrum of women's health topics.

ADAM.com: Women's Health
<http://adam.com/womens_health/womens_health.htm>

Amazingly comprehensive, this site includes detailed information on women's health topics, anatomical images and illustrations of surgical procedures, news items, and information on tests and procedures. It also includes discussion boards on a variety of women's health topics.

allHealth: Women's Health
<http://www.allhealth.com/womens/>

From iVillage.com, this site includes an "Ask the Expert" feature, a virtual checkup that includes visual examples for breast exam, and an osteoporosis risk profile. Health topics include general information along with questions and answers.

AMA Health Insight: Women's Health
<http://www.ama-assn.org/insight/h_focus/wom_hlth/wom_hlth.htm>

Consumer information on women's health topics is provided from the American Medical Association.

drkoop.com: Family Health: Women's Health
<http://www.drkoop.com/family/womens/>

Along with general information on women's health topics, this site includes a question and answer column, weekly essays on topics of interest to women, and chat rooms. Users may sign up for a weekly e-mail newsletter and may personalize the drkoop.com page.

Estronaut: Boldly Exploring Women's Health
<http://www.estronaut.com/index.htm>

A "forum for women's health," this site includes original articles on a variety of women's health topics.

HealthGate: Women's Health
<http://www3.healthgate.com/womenshealth/index.asp>

This site includes featured weekly articles, news, health calculators, and a dictionary.

HealthyWay: Women's Health
<http://sympatico.healthcentral.ca/Centres/OneCentre.cfm?Center= HW%5FWomensHealth>

From Canada, this site offers Canadian health news and links to other Canadian health sites of interest to women. It includes original articles, as well as information from the adam.com medical encyclopedia. It has a search feature.

InteliHealth: Women's Health
<http://www.intelihealth.com/>

The "Women's Health" section of this site by Johns Hopkins includes featured articles, "Ask the Doc," an e-mail health newsletter, and current news. It includes a search engine.

JAMA Women's Health Information Center
<http://www.ama-assn.org/special/womh/womh.htm>

Slanted more toward health professionals, this site includes current news, journal articles, and some patient education resources. There are two special centers, "Contraception Information" and "STD Information."

Mayo Clinic Women's Health Center
<http://www.mayohealth.org/mayo/common/htm/womenpg.htm>

This Mayo Clinic site includes headlines, health quizzes, ask a Mayo physician and reference articles. It also has a search option to locate specific women's health subjects. A unique feature of the Mayo site is an audio dictionary that pronounces medical terms.

The Merck Manual, Home Edition: Women's Health Issues
<http://www.merck.com/pubs/mmanual_home/sec22.htm>

This Web version of *The Merck Manual–Home Edition* is the complete section 22, which includes a variety of women's health issues, and is an excellent source for overview information. There are also links to other pertinent sections of the book. Illustrations are included. This is an excellent source for overview information.

The New York Times: Women's Health
<http://www.nytimes.com/specials/women/whome/index.html>

This site includes articles from *The New York Times* as well as links to other sites. One feature is an annotated guide to over 100 Web sites on women's health.

OBGYN.net: The Obstetrics & Gynecology Network
<http://www.obgyn.net/>

This site includes areas for health practitioners, consumers, and the medical industry. Although sometimes hard to navigate, it contains a large variety of articles on many women's health topics. It also contains information in both Spanish and Portugese.

OnHealth: Women's Health
<http://www.onhealth.com/ch1/resource/conditions/sub7.asp>

OnHealth contains a list of women's health topics, A-Z, and "Ask Our Experts" section, as well as live talk shows from Cleveland Clinic and surgery shows from Stanford University Medical Center. Transcripts of previous shows are also available at this site.

WebMD–Healthy Women
<http://my.webmd.com/living_better/her>

Along with featured articles, the site provides chat capability, message boards, news, and an "Ask the Experts" section. Information on specific health topics is available through a search engine.

WellnessWeb: Women's Health Center
<http://www.wellweb.com/WOMEN/women.htm>

WellnessWeb offers a unique site that includes e-mail discussion alongside specific articles on women's health. Another feature is the "Surgery Center" that discusses pre- and post-operative care, as well as illustrations of the surgical procedure.

Women's Health
<http://womenshealth.miningco.com/health/womenshealth/>

Along with original articles, this site includes links to other health information providers, such as Mayo Clinic and adam.com. It includes book reviews on women's health books, a bulletin board, chat rooms, and a free weekly e-mail newsletter.

Women's Health Interactive
<http://www.womens-health.com/>

A special feature of this site is its interactivity. There are quizzes on a variety of topics that help diagnose health problems. The five health centers–"Headache Center," "Gynecologic Health Center," "Midlife Health Center," "Mental Health Center," and "Natural Health Center"–guide the reader through an interactive learning process.

Women's Health: mylifepath.com
<http://www.mylifepath.com/topic/womens>

Produced by Blue Shield of California, this site includes links to magazine and journal articles, a news archive, health quizzes, and "Ask the Expert." Articles on specific topics are detailed, with links to further information sources.

Women's Health–Natural & Alternative Approaches–HealthWorld Online
<http://www.healthy.net/womenshealth/>

This site has information on a variety of alternative and complementary medicines in women's health, including pages on "Integrative Medicine and Women's Health," "Naturopathic Medicine and Women's Health," "Herbs and Women's Health," "Homeopathy and Women's Health," "Nutrition and Women's Health," and "Acupuncture." It also includes checklists to use when calling your doctor or for doctor visits.

WOMEN'S HEALTH "WEBLIOGRAPHIES"

Britannica.com: Women's Health
<http://www.britannica.com/>

By selecting "The Web's Best Sites" from the health category and then selecting "Women's Health," a user can receive a rating of women's health sites on the Web with annotations. General and specific topics are represented.

Feminist.com: Women's Health
<http://www.feminist.com/health.htm>

Links to general women's health sites, as well as sites on women's cancers, reproductive health, and women and AIDS are available on this site.

Hardin MD: Obstetrics, Gynecology & Women's Health
<http://www.lib.uiowa.edu/hardin/md/obgyn.html>

With the motto, "We list the *best* sites that list the sites," this site has a list of large, medium, and short lists of links to women's health information.

HealthWeb: Women's Health
<http://www.medsch.wisc.edu/chslib/hw/womens/>

This site is slanted more at the health professional than the consumer and contains separate lists for general and specific links to women's health.

MEL: Women's Health
<http://mel.lib.mi.us/health/health-women.html>

The Michigan Electronic Library provides a list of sites aimed at consumers.

The *New York Times* Women's Health: Resources
<http://www.nytimes.com/specials/women/whome/resources.html>

With over 100 annotated Web sites on women's health, including both general and subject specific sites, the *New York Times* offers quality information.

Oregon Health Sciences University Center for Women's Health
<http://ww.ohsu.edu/women>

This site includes links to a variety of resources for women's health, as well as a young women's health page. A unique feature is information on regional women's health conferences and talks.

Women's Health Information Resources
<http://cpmcnet.columbia.edu/dept/rosenthal/Women.html>

From the Rosenthal Center for Complementary and Alternative Medicine, this site presents a list of complementary and alternative medicine resources, as well as specific sources for women's health.

Women's Health Links
<http://som1.umaryland.edu/womenshealth/links/healthlinks.html>

Women's Health Links offers a comprehensive list of links to specific conditions and diseases in women's health from the University of Maryland, and includes links to organizations.

SEARCH ENGINES WITH WOMEN'S HEALTH LINKS

The following sites are from search engine directories. The name of the search engine and URL of the site are included, along with the method of getting to the list in the directory.

AltaVista
<http://dir.altavista.com/Health/47210/47929.shtml>

AltaVista : Health : By Age and Gender : Women's Health

Excite
<http://www.excite.com/health/family_health/for_women/>

Excite : Health : Family Health : For Women

HotBot
<http://dir.hotbot.lycos.com/Health/By_Age_and_Gender/Womens_Health/>

HotBot : Health : By Age and Gender : Women's Health

Infoseek
<http://infoseek.go.com/Center/Health/Family_health/Womens_health?svx=
lhs_ women>

Infoseek : Health : Women's Health

Lycos
<http://dir.lycos.com/Health/Womens_Health/>

Lycos : Health : Women's Health

Snap
<http://home.snap.com/directory/category/0,16,-1529,00.html>

Snap : Health : Women's Health

GOVERNMENT RESOURCES

The following resources are from government sources. Most have full-text health information included, while many of them have statistics and research information included as well.

Canadian Women's Health Network (CHWIN)
<http://www.cwhn.ca/indexeng.html>

This site includes information on women's health in Canada, along with links to women's health resources, some in French.

CDC Health Topic: Women's Health
<http://www.cdc.gov/health/womensmenu.htm>

A variety of health information from the Centers for Disease Control and Prevention is available on this site, including statistics, reports, and pamphlets.

CDC/OD Office of Women's Health
<http://www.cdc.gov/od/owh/whhome.htm>

This site provides women's health information on violence and injury, sexually transmitted diseases, HIV/AIDS, tobacco use, health in later years, reproductive health, and breast and cervical cancer from the Centers for Disease Control and Prevention.

FDA Office of Women's Health
<http://www.fda.gov/womens/default.htm>

This Federal Drug Administration site includes information on studies in women's health, as well as full-text articles and links to other pertinent sites.

National Women's Health Information Center
<http://www.4woman.gov/index.htm>

Health topics information, current events, and health information for special groups comprise some of the links on this site. There is also a special section for Spanish language information.

Office of Research on Women's Health
<http://www4.od.nih.gov/orwh/>

From an Office of the National Institutes of Health, this site includes information on women's health research, women in biomedical careers, and women as participants in research.

Office on Women's Health, Department of Health and Human Services
<http://www.4woman.gov/owh/>

This site includes online publications, links to Centers of Excellence, information on women's programs, as well as a link to the National Women's Health Information Center.

WHO–Women's Health and Development
<http://www.who.int/frh-whd/index.html>

The WHO site contains information on international women's health issues with links to other sites.

ORGANIZATION SITES

These organizations have information on women's health, some available in full text.

ACOG–American College of Obstetricians and Gynecologists
<http://www.acog.org/>

The ACOG site includes a physician directory, patient education, and information on women's health issues.

The American Medical Women's Association (AMWA)
<http://amwa-doc.org/>

AMWA is an organization of 10,000 women physicians and medical students dedicated to serving as the unique voice for women's health and the advancement of women in medicine. The site includes excerpts from *The Women's Complete Healthbook* published by AMWA.

Association of Professors of Gynecology and Obstetrics (APGO)
<http://www.apgo.org/>

With information on teaching women's health for professors and students, this site includes curriculum models.

Boston Women's Health Book Collective
<http://www.ourbodiesourselves.org/Default.htm>

From the authors of *Our Bodies, Ourselves,* this site includes excerpts from the book, as well as links to a variety of other resources.

familydoctor.org
<http://familydoctor.org/>

Offering health information from the American Academy of Family Physicians, this site includes information on a number of women's health topics.

International Women's Health Coalition
<http://www.iwhc.org/>

This site focuses on information on women's health issues around the world.

Jacobs Institute of Women's Health
<http://www.jiwh.org/index.htm>

Jacobs, a not-for-profit organization dedicated to advancing knowledge and practice in the field of women's health, has a site that includes links to other resources.

National Women's Health Network
<http://www.womenshealthnetwork.org/>

The NWHN is a nonprofit health advocacy organization founded in 1975 to give women a greater voice in the health care system in the United States. The Network advocates better federal policy on women's health and through its Information Clearinghouse, provides women with information and resources to assist them in making better health care decisions.

North American Menopause Society
<http://www.menopause.org/>

Although mostly concerned with menopause issues, this site also has links to a select group of other sites, as well as full-text information.

Society for Women's Health Research
<http://www.womens-health.org/>

The Society for Women's Health Research seeks to improve the health of women through research. This site includes links to other resources on research as well as statistics in women's health research.

Women's Health Project
<http://www.whealth.org/>

The Women's Health Project is a consortium of nine institutions that have developed a traveling exhibit, *The Changing Face of Women's Health.* This site includes some of the interactive parts of the exhibit, as well as teachers' resources.

NEWS AND E-MAIL LIST SITES

CNN.com–Health : Women
<http://www.cnn.com/HEALTH/women/>

This site features current news on women's health from CNN.

Medscape Women's Health
<http://www.medscape.com/Home/Topics/WomensHealth/womenshealth.html>

The Medscape Women's Health site provides current health information, with a weekly update subscription capability.

2000 Daily News
<http://www.4woman.gov/nwhic/News/index.htm>

From the National Women's Health Information Center, the daily news site focuses on women's health.

WebMedLit: Women's Health
<http://webmedlit.silverplatter.com/topics/womens.html>

WebMedLit has a list of current articles on women's health, with e-mail updates available.

E-mail Lists About Women's Health
<http://research.umbc.edu/~korenman/wmst/f_hlth.html>

E-mail lists focus on women's health issues. Subscription information is included.

YOUNG WOMEN'S HEALTH

Young women from preteen through early adulthood have unique health information needs, including eating disorders, puberty, and the first gynecological exam. The following sites all have information of interest to these young women.

Femina: Girls' Health
<http://femina.cybergrrl.com/netscape.htmealth>

Under the subheadings of "Girl" and "Health," there are a variety of links on this site, including eating disorders, menstruation, and AIDS.

Girl Power! Get Body Wise!
<http://www.health.org/gpower/girlarea/bodywise/index.htm>

Girl Power contains information on fitness, nutrition, body image, and eating disorders.

GirlSpace
<http://www.kotex.com/girlspace/>

From Kotex, this site includes information about menstruation, as well as a first visit to a gynecologist.

gyn 101
<http://www.gyn101.com/homefr.htm>

Gyn 101 has information on a gynecological exam, including what will happen, what questions to ask, and a self-test.

Information for Young Women
<http://www.womanhealth.net/html/youngwomen.html>

With original articles on a number of health topics for young women, this site also provides links to other pertinent sites.

OBGYN.net–Young Woman Section
<http://www.obgyn.net/yw/yw.htm>

Although OBGYN.net is sometimes hard to navigate, it includes a variety of articles of interest to young women.

Student Advantage: Women's Health
<http://www.studentadvantage.com/issue/1,1061,c5-i50,00.html>

A site aimed at the college-age student, it includes information on sexual health, eating disorders, and nutrition.

Teen Health Web Site
<http://www.chebucto.ns.ca/Health/TeenHealth/index.html>

Produced by Dalhousie Medical School, the Teen Health Web Site includes information on sexual orientation, sexual assault, pregnancy, and specific sites for women's health.

Teen Health–Women's Health Net Links
<http://womenshealth.miningco.com/health/womenshealth/msubteen.htm>

This site provides links to original articles and to other Web sites with information on a variety of young women's health topics.

Young Women's Health Page
<http://www.ohsu.edu/women/teen>

Produced at the Oregon Health Sciences University Center for Women's Health by high school students, this site includes information on physical health, mental health, and sexuality for young women, as well as hotline listings.

Young Women's Resource Center Youth Page
<http://www.youngwomenshealth.org/youthpage.html>

Produced at the Children's Hospital in Boston, this site includes a list of recommended books and Web sites for young women.

SENIOR WOMEN'S HEALTH

Women in their later years have a variety of health information needs unique to their age group. Some of these include menopause, osteoporosis, sexuality, and diseases that have aspects unique to the older woman. These sites contain information specifically for this age group.

AARP Webplace: Explore Health
<http://www.aarp.org/healthguide/home.html>

Although not specifically for women, this American Association of Retired People site contains a great deal of health information for senior women.

InteliHealth: Women and Aging
<http://www.intelihealth.com/>

A feature in the Women's Health section, the InteliHealth site contains information from Johns Hopkins on topics such as Alzheimer's, menopause, sexuality, and osteoporosis.

National Center on Women and Aging
<http://www.heller.brandeis.edu/national/ind.html>

This site contains excerpts from the *Women and Aging* newsletter, as well as links to other sites.

OWL: National Older Women's League
<http://www.owl-national.org/>

OWL is a "national membership organization that seeks to improve the status and quality of life of midlife and older women." The site has information on heart disease and osteoporosis, as well as gender inequality issues, social security, and Medicare facts.

Senior Women Web: Health and Fitness
<http://www.seniorwomen.com/healthfitness.htm>

The Senior Women Web site provides links to a variety of sites on health and fitness.

ThirdAge–Women's Health
<http://www.thirdage.com/guides/health/women/>

The ThirdAge site provides information on a variety of women's health topics, as well as alternative medicine information, and includes a health forum for questions, answers, and comments.

Women's Health-Senior Health
<http://seniorhealth.about.com/health/seniorhealth/msubwomenhealth.htm?
once=true&>

Women's Health features information on balance disorders, menopause, osteoporosis, and a variety of other topics.

SPECIFIC GROUPS

These Web sites spotlight specific groups of women: ethnic, lesbian, disabled, and rural women. They contain information that may have a greater impact for these groups or health aspects that are of interest only to these specific groups. Some of them also have health statistics.

African American Women's Health
<http://www.4woman.gov/faq/african_american.htm>

From the National Women's Health Information Center, this site features information on health issues of particular interest to African American women.

Asian and Pacific Islander Women's Health
<http://www.4woman.gov/faq/Asian_Pacific.htm>

From the National Women's Health Information Center, this site includes information on health issues of particular interest to Asian and Pacific Islander women.

Black Health Net: Women's Health
<http://www.blackhealthnet.com/>

The "Women's Health" section of this page contains original articles on a variety of health issues for African American women.

BlackWomensHealth.com
<http://www.blackwomenshealth.com/>

With information on health issues for African American women, this site features original articles and statistical information.

Health Information Resources for Rural Women
<http://www.nal.usda.gov/ric/richs/women.htm>

From the Rural Information Center Health Service, this site provides a variety of health links of interest to women living in rural areas.

Latina Women's Health
<http://www.4woman.gov/faq/latina.htm>

From the National Women's Health Information Center, this site includes information on health issues of particular interest to Latina women.

Lesbian Health
<http://www.lesbian.org/lesbian-moms/who.html>

This site gives a variety of links on lesbian health concerns, including pregnancy, parenting, and legal issues.

National Asian Women's Health Organization
<http://www.nawho.org/index.html>

NAWHO was formed to improve the health status of Asian American women and families through research, education, and public policy advocacy. The site includes information about the various programs in which it is involved.

Native American Women's Health
<http://www.4woman.gov/faq/native_american.http>

From the National Women's Health Information Center, this site provides information on health issues of particular interest to Native American women.

Women and Disability Resources
<http://members.tripod.com/~Barbara_Robertson/Women.html>

This site has information about women and disabilities, along with links to a variety of sources.

Women of Color Health Data Book
<http://www.4women.gov/owh/pub/woc/index.htm>

From the National Women's Health Information Center, this site contains information and statistics on all aspects of health for all women of color in the United States.

Women with Disabilities
<http://www.4women.gov/wwd/index.htm>

From the National Women's Health Information Center, the Women with Disabilities site features information on all aspects of disability from financial to social. It also includes links for information on specific disabilities.

CONCLUSION

There are, of course, so many women's health Web sites that it is impossible to include all in a list. This is an overview of some of the best sites, and with the Internet changing rapidly, some may have changed radically or no longer be in existence by the time this paper is published.

Good health information for women is something that was not readily available, even within the last decade. With the information currently available and the growth of women using the Internet, we can hope that women will have more control over their health in the future.

Evaluating and Choosing Web Sites on Women's Health: The Perspective from Academe

Ellen Gay Detlefsen
Nancy Hrinya Tannery

ABSTRACT. This article, in two parts, presents information on the evaluation of Web sites in general and the choice of women's health Web sites in particular. A review of studies and tools for Web site evaluation is followed by a core list of women's health Web sites, from the perspective of the academic health sciences library. *[Article copies available for a fee from The Haworth Document Delivery Service: 1-800-342-9678. E-mail address: <getinfo@haworthpressinc.com> Website: <http://www.HaworthPress.com>]*

KEYWORDS. Women's health, Web site evaluation

INTRODUCTION

According to the National Institutes of Health, women make 60% of all physician visits and spend one out of every three health care dollars.[1] It

Ellen Gay Detlefsen, DLS (ellen@sis.pitt.edu), is Associate Professor, Department of Library and Information Science, School of Information Sciences, University of Pittsburgh, 135 N. Bellefield Avenue, 651 LIS Building, Pittsburgh, PA 15260. She is also Associate Professor in the Women's Studies Program at the University, a member of the Core Training Faculty of the Center for Biomedical Informatics, University of Pittsburgh School of Medicine, and Project Director for the Highmark Minority Health Link. Nancy Hrinya Tannery, MLIS (nht2120+@pitt.edu), is Assistant Director for Information Services at the Health Sciences Library System, University of Pittsburgh, and Web Coordinator for the HSLS/Falk Library.

[Haworth co-indexing entry note]: "Evaluating and Choosing Web Sites on Women's Health: The Perspective from Academe." Detlefsen, Ellen Gay, and Nancy Hrinya Tannery. Co-published simultaneously in *Health Care on the Internet* (The Haworth Press, Inc.) Vol. 4, No. 2/3, 2000, pp. 19-31; and: *Women's Health on the Internet* (ed: M. Sandra Wood, and Janet M. Coggan) The Haworth Press, Inc., 2000, pp. 19-31. Single or multiple copies of this article are available for a fee from The Haworth Document Delivery Service [1-800-342-9678, 9:00 a.m. - 5:00 p.m. (EST). E-mail address: getinfo@haworthpressinc.com].

follows that they would be interested in locating health information on the Web. Market research studies on Internet users have identified the specific phenomenon of the woman who is a "med-retriever."[2-3] Enter the terms "women's health" in the search engine AltaVista and a user would retrieve approximately 190,000 Web pages.[4] How could anyone, physician, nurse, student, or consumer, wade through all these thousands of Web pages and find useful, authoritative information?

This article has two parts: the first section provides background information and tips on the evaluation of Web sites in general, and the evaluation of Web sites that focus on women's health in particular, and the second section provides a basic or core list of women's health resources that can be used as the foundation for an academic health sciences library pathfinder or Web-based bibliography.

PART ONE: EVALUATING WEB SITES IN GENERAL

The literature–both print and Web-based–of library/information science and of medicine is already replete with articles and tip sheets on the evaluation of Web-based information. Studies "rating" everything from health mega-sites to individual Web pages on very specific topics are being published almost weekly. Most are generic, however, and cover classic "markers" to look for when reviewing Web sites and Web pages.[5-12] One of the most thorough of the general studies is the health megasite evaluation project, spearheaded by a team from the University of Michigan libraries; it contains both an extended discussion of general evaluation criteria, an excellent bibliography, and a handy checklist.[13]

Certain items appear so often in these studies and reports that they can be listed as standards; those that appear in the "Selection Guidelines" for the National Library of Medicine's MEDLINE*plus* are typical. The NLM lists these standard attributes as the following.

Quality, authority and accuracy of content. The source of the content is established, respected and dependable. A list of advisory board members or consultants is published on the site. The information provided is appropriate to the audience level, well-organized, and easy to use. Information is from primary resources (i.e., textual material, abstracts, Web pages). Lists of links are evaluated/reviewed/quality-filtered.

The purpose of the Web page is educational and is not selling a product or service. Most content is available at no charge.

Availability and maintenance of the Web page. The Web site is consistently available. Links from the site are maintained. The source for the contents of the Web page(s) and the entity responsible for maintaining the Web site (webmaster, organization, creator of the content) is clear. Information is current or an update date is included.

Special features. The site provides unique information to the topic with a minimum of redundancy and overlap between resources. The site contains special features such as graphics/diagrams, glossary, or other unique information. The content of the site is accessible to persons with disabilities.[14]

The Greater Midwest Regional Library of the National Network of Libraries of Medicine has a typical tip sheet, which identifies important evaluation criteria as "Content, Audience, Authority/Source, Date/Timeliness, [and] Structure/Access."[15] The Medical Library Association, in an effort to help educate the general public about the issues they–the public–face when seeking health information on the World Wide Web, put out their guide to "Diagnosing Web Sites," including a short series of cautionary questions and ending with some advice:

- Who sponsors the Web site?
- Is the site current?
- Has it been updated recently?
- Is the information factual or does it represent opinions?
- Is the Web site intended for medical professionals or the general public?
- Need more information? Try asking a medical librarian . . . [16]

EVALUATION RATINGS

The phenomenon of organized evaluative rating of Web sites, and the awarding of icons or banners to sites rated as excellent or useful, also seems well established. Sometimes labeled "intermediary tools," these evaluations are now widespread and popular as markers for quality; Web sites and their creators cherish their awards, and present them prominently on home pages.[17] Among the most widely sought after of such designations in the medical field is the imprimatur of the international Health on the Net Foundation, which specifically lists eight principles on which it rates candidate Web sites which present medical and health information. These principles are:

- Authority
- Complementarity
- Confidentiality
- Attribution
- Justifiability
- Transparency of Authorship
- Transparency of Sponsorship
- Honesty in Advertising and Editorial Policy[18]

Another highly-regarded award is the "Hardin MD Clean Bill of Health Award," a service from the University of Iowa's Hardin Library of the Health Sciences that boasts "We list the best sites that list the sites," and awards its logo to health Web sites that are "among the best in comparison with all link lists on the net," specifically those whose links have "connection rates of at least 93%."[19]

Another evaluative honor is selection by a site as the "website of the week" by the *British Medical Journal,* a form of ongoing peer review for health Web sites.[20] A sort of reverse award exists in the designation of "bad" Web pages, where the raters and evaluators are publicizing poor design, bad links, outdated and inaccurate information, etc. These awards are seldom advertised by the offending Web sites![21]

EVALUATING WEB SITES FOR SPECIFIC POPULATIONS

While these generic lists of attributes, markers, and evaluation guidelines are good, they usually do not take into account the special needs of different populations–both populations of users *and* the populations being addressed by the Web sites and pages themselves. There are well-developed guidelines that discuss the evaluation of Web sites from the perspective of a few specialized user groups; most are targeted to the undergraduate in colleges and universities.

A tutorial, developed at the University of California at Berkeley, is typical of this approach. It is part of a two-part Internet Workshop offered year-round by the team of teaching librarians who staff the university library's Internet Instruction Program. One helpful feature of this vast tutorial is a "worksheet for analyzing your topic" that helps those doing Web-based searches to frame their queries more usefully.[22] Another classic from this "bibliographic instruction" approach is Esther Grassian's "Thinking Critically About World Wide Web Resources," which succinctly presents key questions to ask in the areas of "Content & Evaluation," "Source & Date," "Structure," and "Other." Grassian also recommends a companion page entitled "Thinking Critically about Discipline-Based World Wide Web Resources," which takes a more specific approach to framing Web searches and evaluating Web retrieval from a subject point of view.[23] One additional Web evaluation tool is that of Jim Kapoun, a reference and instruction librarian at Southwest State University. This site, published under the auspices of the ALA's Association of College and Research Libraries, contains a handy chart of tips for evaluating Web-based information in the classic areas of accuracy, authority, objectivity, currency, and coverage. Kapoun's chart contains two columns: "Evaluation of Web Documents" on the left, and "How to Interpret the Basics" on the right; it is ideal for pinning up on a bulletin board or duplicating as a cover sheet for evaluation studies.[24]

It is far more rare to find tools, or even research studies, which frankly provide information or advice on the evaluation of Web sites that cover information from the perspective of a specific population other than undergraduate students. This is probably the result of the paucity of research on the information behaviors of various groups; most research work studying information behavior in the health sciences has been focused on the information behaviors of physicians, medical students, and nurses.[25]

There is some evidence of work on the issue of designing and evaluating Web sites for those with limited reading skills, those who are functionally illiterate. One set of researchers has recently investigated the readability levels of Web-based consumer health information.[26] A handful of studies have appeared in the medical literature, reporting studies by physicians who have evaluated specific subject-oriented Web sites as tools for parents seeking health information about their children.[27-28] An article in 1997 specifically analyzed the Web-based resources in men's health, prepared by Jerry Perry, a reference librarian now at the University of Arizona.[29]

Possibly the most useful model for evaluating Web-based information from the perspective of a specific population is the superb "Techniques for Evaluating American Indian Web Sites," created by Elaine M. Cubbins and available through the University of Arizona. This is a far more sophisticated evaluation tool than most, in that it explicitly describes the role that the assessment of the Native American content plays in the evaluation process and offers advice on how to evaluate that content.[30]

A handful of existing sites and studies are available that focus specifically on evaluation of women's health. An early research study appeared in the *Journal of Women's Health* in 1997,[31] while a news story approach was featured in a 1999 CNN.com feature. In the latter piece, after a discussion of several common urban myths in women's health, the article ends with a few words of advice to consumers who do their own Web-based health research.[32] Finally, a consumer-oriented hot-linked list of resources for women's health can be found on the excellent Web site maintained by the Consumer and Patient Health Information Section (CAPHIS) of the Medical Library Association; this list was compiled by an experienced consumer health librarian from Michigan.[33]

ISSUES IN THE EVALUATION
OF WOMEN'S HEALTH WEB SITES

The issues of evaluating women's health Web sites are complex. Those who do such evaluations–for themselves or as collections development specialists for libraries–must consider two sets of criteria. In addition to looking for the presence of the classic markers and attributes, for the awards of evaluative ratings by outside reviewers, and for the reviews in research studies, evaluators must ask the following additional questions:

- Does the site produced explicitly identify itself as a women-centered organization of individual? Is it created and hosted by a women's clinic or women's health specialist?
- Does the site address both those health concerns specific to women only, as well as general health issues that are known to occur in female populations?
- Are the issues of the various women's sub-populations acknowledged and addressed? Is there coverage of girls, teen women, and seniors? How are the needs of lesbians, women of color, women with disabilities, poor women, women with mental illness, and women who are the victims of violence handled?
- Are the images and pictures on the site of women and girls, both as the patients/consumers and as health professionals?
- How does the site deal with political issues or issues of sexuality? Is there information about access to abortion and contraception? Is there information about legal rights with respect to health insurance or access to health care? Is the site neutral, or does it have a point of view?
- Does the site explicitly address the health information needs of women who may not themselves be the patient/consumer, but may be the family caregiver? Does it speak to the needs of women as health professionals?

In choosing women's health Web sites to recommend, the combination of an awareness of these kinds of women-centered concerns together with the more traditional measures of excellence, will result in a list of useful sites that can serve as the building blocks for an Internet resource guide or Web bibliography.

PART TWO: A BASIC LIST OF WOMEN'S HEALTH WEB SITES

This core list is not meant to be exhaustive, but is rather a recommended list of starting points for locating women's health information. Nine general sites and six more specific sites are annotated; other useful sites are simply listed. All the links described in this article can be found on the University of Pittsburgh's Health Sciences Library System Web site, at <http://www.hsls. pitt.edu/intres/health/women.html>.

General Women's Health Web Sites

The National Women's Health Information Center (NWHIC)
<http://www.4woman.gov/>

NWHIC is project of the Office on Women's Health in the Department of Health and Human Services. This Web site links the user to government resources directed towards all aspects of women's health. Daily news stories and dictionaries and glossaries of health terms are included in this site.

Centers for Disease Control and Prevention (CDC)
<http://www.cdc.gov/health/womensmenu.htm>

The CDC provides links to fact sheets and documents covering a broad range of women's health topics. Information can be located on assisted reproductive technology, folic acid, unintended pregnancy, and about 15 more topics. Links to full-text articles in journals such as the *New England Journal of Medicine* and government documents are included.

Office of Research on Women's Health (ORWH)
<http://www4.od.nih.gov/orwh/>

ORWH is a component of NIH whose goal is to increase research in women's health. It ensures that women are included in all human subject research. The Web site provides links to research, publications, and meetings and seminars. Included in this site is an extensive list of consumer health Web sites dealing with various women's health issues.

The *JAMA* Women's Health Information Center
<http://www.ama-assn.org/special/womh/womh.htm>

This Web site is produced and maintained by the *Journal of the American Medical Association's (JAMA)* editors and staff and is developed as a resource for physicians and health care professionals. The "Library" link is a journal scan of current peer reviewed full-text journal articles about women's health. This list of journal articles is updated weekly. The "Newsline" feature links users to weekly news articles about women's health, special reports, conference coverage, and background briefings. There are two in-depth information sections included in this *JAMA* Web site, one on sexually transmitted diseases, and the other on contraception information.

The American Medical Women's Association (AMWA)
<http://www.amwa-doc.org/>

The AMWA is an organization of women physicians and women medical students who want to serve as a voice for women's health. The Web site provides links to education initiatives, advocacy issues, the *Journal of the American Medical Women's Association,* and other publications. It provides consumer health topics of interest to women including information on thyroid disease, cervical cancer, and aging.

Society for Women's Health Research
<http://www.womens-health.org/>

This professional society seeks to improve women's health through re-search. Emphasis is placed on gender-based biology that tries to uncover the biological and physiological differences between men and women. The site provides results of research in women's health, links to policy and advocacy information, clinical trials, and research funding. Links to health sites and facts of interest to women's health are also included.

Women's Health Topics, MEDLINE*plus*
<http://www.nlm.nih.gov/medlineplus/womenshealth.html>

Women's Health Topics contains health care information compiled by the National Library of Medicine. The site links to government and non-govern-ment publications, brochures, databases, and Web sites. More than 40 topics are covered including breast and ovarian cancer, infertility, pregnancy, and vaginal diseases. Each topic links to a search in PubMed, NLM's free Web-based MEDLINE. Health professionals and consumers will find useful, up-dated information from this Web site.

National Women's Health Resource Center
<http://www.healthywomen.org/>

Supplying health care consumers with women's health information is the goal of the National Women's Health Resource Center (NWHRC). A ques-tion and answer segment and links to topical Web sites provide users with valuable information on a variety of women's health topics. These topics cover depression, diabetes, osteoporosis, and thyroid disease to name a few. Membership is an option but information is accessible to all users of the site. Professional searches on personal health topics are available, but there is a fee for nonmembers.

Boston Women's Health Book Collective
<http://www.ourbodiesourselves.org>

As a well-known and award-winning publisher of the classic, *Our Bodies, Ourselves for the New Century,* the Collective is "a nonprofit women's health education, advocacy and consulting organization [whose] broad purpose is to help individuals and groups make informed personal and political decisions about health and medical care, especially as they relate to women." The site contains a large number of links to both information on general women's health issues and to specific illnesses that predominantly or solely affect women.

Other General Women's Health Web Sites of Interest

Women's Health in the U.S.: NIAID Research on Health Issues Affecting Women
<http://www.niaid.nih.gov/publications/womenshealth/textonly.htm>

FDA and Women's Health Issues
<http://www.fda.gov/womens/informat.html>

Women's Health and Gender Differences, National Institute on Drug Abuse (NIDA)
<http://www.health.org/links.htm>

A Forum for Women's Health
<http://www.womenshealth.org/>

Specific Women's Health Web Sites

Office of Minority and Women's Health (OMWH)
<http://www.bphc.hrsa.gov/omwh/omwh.htm>

OMWH is an office of the Health Resources and Services Administration (HRSA). This site provides information on health-related issues and data trends specifically focused on minority women's health. Areas covered include breast cancer, domestic violence and older women.

Women of Color Health Data Book
<http://www.4women.gov/owh/pub/woc/index.htm>

The Women of Color Health Data Book covers the health of four major groups of women, Native, Hispanic, black, and Asian Americans. Issues covered in this data book are factors affecting the health of women of color, health assessment of women of color, and improving the health of women of color.

Older Women: National Aging Information Center
<http://www.aoa.gov/NAIC/Notes/olderwomen.html>

The National Aging Information Center is a service of the Administration on Aging. Listed here are Web sites that provide links to statistics, organizations, fact sheets, and research and academic sites that provide health information geared toward the older woman.

On the Teen Scene
<http://www.fda.gov/oc/opacom/kids/html/7teens.htm>

The Food and Drug Administration is responsible for the information on this Web site. Although not all the information is directed towards teenage girls, this Web site contains plenty of useful information. Links to documents with information about menstruation, cosmetics, eating disorders, Toxic Shock Syndrome, and yeast infections can be found here. The documents contain images, diagrams, references, and links to associations.

Adolescent Pregnancy and Reproductive Health
<http://www.hsph.harvard.edu/Organizations/healthnet/maternal/topic23.html>

Global Reproductive Health Forum @ Harvard (GRHF), an endeavor undertaken by the Harvard School of Public Health, has developed a Web site that provides links to information on adolescent reproductive health. Guides and Information, Data, Trends, and Statistics, Peer Education and Information, Organizations, Activities and Associations, and Adolescent Rights are the categories of information that can be found at this Web site.

Women with DisAbilities
<http://www.4women.gov/wwd/index.htm>

This Web site, compiled by the National Women's Health Information Center (NWHIC), is intended to provide health information for women with disabilities or women who care for a person(s) with a disability. General resources about critical health issues for a variety of disabilities are provided, including physical, neurological, hearing, speech, and visual impairment. It will also provide information on psychiatric, learning, and developmental disabilities. The health issues addressed include sexuality, abuse, parenting, and substance abuse.

Other Specific Women's Health Web Sites of Interest

The Many Dimensions of Depression in Women: Women at Risk,
National Institute of Mental Health (NIMH)
<http://www.nimh.nih.gov/depression/women/risk.htm>

Environews by Topic Women's Health, National Institute of
Environmental Health Sciences (NIEHS)
<http://ehis.niehs.nih.gov/topic/women.html>

American College of Obstetricians and Gynecologists
<http://www.acog.org/>

CONCLUSION

This article provides focused advice on the evaluation of Web sites with a specific emphasis on women's issues, and a sample of the types and variety of women's health resources available on the Web. These sites can be used as the foundation for any library's guide to Web-based resources on women's health, both for those interested in general women's health sites as well as pointers to sites directed towards more specific women's health needs and sub-populations.

NOTES

1. Dennis-Smith, R. "Why Health Plans Should Court Female Members." *Employee Benefits News* 13(5, 1999):42-3.

2. Health on the Net Foundation's 5th HON Survey on the Evolution of Internet Use for Health Purposes (October-November 1999). Available: <http://www.hon.ch/ Survey/ResultsSummary_oct_nov99.html>.

3. Georgia Institute of Technology Graphic, Visualization, & Usability Center's 10th WWW User Survey (October 1998). Available: <http://www.cc.gatech.edu/ gvu/user_surveys/survey-1998-10/graphs/use/q109.htm>.

4. AltaVista [search engine online]. Palo Alto, Calif: AltaVista Co. Available: <http://www.altavista.com/>. Accessed: December 21, 1999.

5. Coiera E. "The Internet's Challenge to Health Care Provision." *British Medical Journal* 312(1996):3-4. Available: <http://www.bmj.com/cgi/content/full/312/ 7022/3?ijkey=beSS5LxxK5yqw/>.

6. Jadad, A.R., and Gagliardi, A. "Rating Health Information on the Internet. Navigating to Knowledge or to Babel?" *JAMA* 279(February 25, 1998):611-4.

7. Kim, P.; Eng, T.R.; Deering, M.J.; and Maxfield, A. "Published Criteria for Evaluating Health Related Web Sites: Review." *British Medical Journal* 318 (1999):647-9. Available: <http://bmj.com/cgi/content/full/318/7184/647>.

8. Lindberg, D.A.B., and Humphreys, B.L. "Medicine and Health on the Internet: The Good, the Bad, and the Ugly." *JAMA* 280(October 21, 1998):1303-4.

9. Mitretek Systems–Health Information Technology Institute [and] U.S. Agency for Health Care Policy and Research. "White Paper: Criteria for Assessing the Quality of Health Information on the Internet." Available: <http://hitiweb.mitretek. org/docs/criteria.html>.

10. Murray, S. "Separating the Wheat from the Chaff: Evaluating Consumer Health Information on the Internet." *Bibliotheca Medica Canadiana* 19(Summer 1998):142-5.

11. Silberg, W.M.; Lundberg, G.D.; and Musacchio, R.A. "Assessing, Controlling, and Assuring the Quality of Medical Information on the Internet." *JAMA* 277(April 16, 1997):1244-5.

12. Wyatt, J.C. "Commentary: Measuring Quality and Impact of the World Wide Web." *British Medical Journal* 314(1997). Available: http://www.bmj.com/archive/ 7098ip2.htm/.

13. Anderson, P.F. et al. "The Megasite Project: A Metasite Comparing Health Information Megasites & Search Engines." Available: <http://www.lib.umich.edu/ megasite/>.

14. "MEDLINE*plus* Selection Guidelines." Available: <http://www.nlm.nih.gov/ medlineplus/criteria.html/>.

15. NN/LM. Greater Midwest Region. "Tips on Evaluating Web Resources." Available: <http://www.nnlm.nlm.nih.gov/gmr/publish/eval.html/>.

16. Medical Library Association. "Medspeak: Diagnosing Web Sites." Available: <http://www.mlanet.org/resources/medspeak/meddiag.html>.

17. As an example, see Rick Mendosa's site for "Online Diabetes Resources," with his award logos visible at <http://www.mendosa.com/faq.htm>.

18. Health on the Net Foundation. HONCode. Available: <http://www/hon.ch/ HONcode/Conduct.html>.

19. Hardin MD Clean Bill of Health Award. Available: <http://www.lib.uiowa.edu/ hardin/md/cbh.html>.

20. As an example, see BMJ Reviews Multimedia WEBSITE OF THE WEEK Available: <http://www.bmj.com/cgi/content/full/318/7176/135/b>.

21. As an example, see Vincent Flanders's "Web Pages that Suck.com," Available: <http://www.webpagesthatsuck.com/>.

22. UC Berkeley Library. "Finding Information on the Internet. A Tutorial. Available: <http://www.lib.berkeley.edu/TeachingLib/Guides/Internet/FindInfo.html>.

23. Grassian, E. "Thinking Critically about World Wide Web Resources." UCLA College Library Instruction. Available: <http://www.library.ucla.edu/libraries/college/ instruct/web/critical.htm>.

24. Kapoun, J. "Teaching Undergrads WEB Evaluation: A Guide for Library Instruction." *College & Research Libraries News*. Available: <http://www.ala.org/acrl/ indwebev.html>.

25. Detlefsen, E.G. "The Information Behaviors of Life and Health Scientists and Health Care Providers: Characteristics of the Research Literature." *Bulletin of the Medical Library Association* 86(July 1998):385-90.

26. Graber, M.A.; Roller, C.M.; and Kaeble, B. "Readability Levels of Patient Education Material on the World Wide Web." *Journal of Family Practice* 48(January 1999):58-61.

27. Impicciatore, P.; Pandolfini, C.; Casella, N.; and Bonati, M. "Reliability of Health Information for the Public on the World Wide Web: Systematic Survey of Advice on Managing Fever in Children at Home." *British Medical Journal* 314 (7098). Available: <http://bmj.com/cgi/content/full/314/7098/1875/>.

28. McClung, H.J.; Murray, R.D.; and Heitlinger, L.A. "The Internet as a Source for Current Patient Information." *Pediatrics* 101(6, 1998). Available: <http://www.pediatrics.org/cgi/content/full/101/6/e2>.

29. Perry, G.J. "A Guy Thing: Consumer-Oriented Men's Health Resources on the World Wide Web." *Health Care on the Internet* 1(2, 1997):3-9.

30. Cubbins, E.M. "Techniques for Evaluating American Indian Web Sites." <http://www.u.arizona.edu/~ecubbins/webcrit.html>.

31. Wootton, J.C. "The Quality of Information on Women's Health on the Internet." *Journal of Women's Health* 6(5, 1997):575-81.

32. "Email Health Scares: Separating the Facts from the Hype." CNN.com/health. Available: <http://www.cnn.com/2000/HEALTH/02/02/health.scares.wmd/index.html>.

33. Aebli, C. "Finding It on the Web: A Guide to Finding and Evaluating Information on Women's Health on the Internet." Available: <http://caphis.njc.org/Finding.html>.

Women's Health on the Web

Ann Boyer

ABSTRACT. The concept of women's health is discussed, along with its origin and its varying interpretations. Some different types of women's health Web sites are outlined, as defined by their sponsorship and purpose, and examples are given. As a specific example, the Women's Health section of HealthWeb is described in greater detail. *[Article copies available for a fee from The Haworth Document Delivery Service: 1-800-342-9678. E-mail address: <getinfo@haworthpressinc.com> Website: <http://www.HaworthPress.com>]*

KEYWORDS. Women's health, HealthWeb, Internet

INTRODUCTION

The Topic of Women's Health

Getting a handle on the big unruly topic of women's health is like trying to hang onto an octopus. There's always a leg out there waving around that you can't quite get a hold of. Women's health encompasses many disparate topics. It cuts across other fields such as gynecology and obstetrics, rheumatic diseases, mental health, and oncology. Depending upon the definition, it may include areas that are not strictly medical, such as wellness, nutrition, and domestic violence. If you visit many large, inclusive women's health Web sites, you realize that there is not a clear-cut definition of what constitutes

Ann Boyer, MS, MLS (boyer@library.wisc.edu), has been a Reference Librarian at the Health Sciences Library, University of Wisconsin-Madison for the past ten years. She recently retired.

[Haworth co-indexing entry note]: "Women's Health on the Web." Boyer, Ann. Co-published simultaneously in *Health Care on the Internet* (The Haworth Press, Inc.) Vol. 4, No. 2/3, 2000, pp. 33-45; and: *Women's Health on the Internet* (ed: M. Sandra Wood, and Janet M. Coggan) The Haworth Press, Inc., 2000, pp. 33-45. Single or multiple copies of this article are available for a fee from The Haworth Document Delivery Service [1-800-342-9678, 9:00 a.m. - 5:00 p.m. (EST). E-mail address: getinfo@haworthpressinc.com].

women's health; various Web authors construe it to include different mixes of topics. On many women's health pages, the viewer will find links to sites that aren't really "women's health," but simply focus on a medical problem shared by men and women.

The Women's Health Movement

In certain respects, the idea of "women's health" is a cultural construct. It emerged out of the women's movement of the 1960s and 1970s, as some women became more political and began to see many of women's needs as under-addressed by the political and medical establishments. There was a sense that only women could really understand some female medical problems. This was the period when the now-famous health book, *Our Bodies, Ourselves,* was written by the Boston Women's Health Book Collective.[1] "Groups of women all over the country met regularly to share experiences regarding health and illness," writes Rebecca Lovell Scott, "to teach each other about self care, to empower themselves in dealing with a medical system they saw as unresponsive to their needs."[2] A decade later, Norma Swenson reports, the National Women's Health Network was founded to

> bring pressure on the government to provide better safeguards to women, and bring a challenge to all those groups whose interest in women's health was primarily economic, so that government policy would be more in line with women's needs . . . The Network thus became a kind of umbrella and the women's health movement then began to become a practical political force . . . [Today] the women's health movement has formed important coalitions and networks and . . . [has] begun to negotiate as equals with many of the forces which create health policy in the U.S. . . .[3]

Out of this crucible of feminist activism, women's health emerged as a recognized field.

AN OVERVIEW OF WOMEN'S HEALTH WEB SITES

Today on the Web, we find many types of women's health sites offering information that varies dramatically in quality, reliability, and completeness. To begin, we will look at some of the various kinds of sites. The number of topics being almost endless, it was decided to examine sites using this criterion: who created them, and for what purpose?

Governmental Sites

Government agency sites are well represented on this topic. The U.S. Department of Health and Human Services (DHHS) Office on Women's Health sponsors the National Women's Health Information Center <http://www.4woman.gov/>, intended for use by diverse groups including consumers, health care professionals, researchers, educators, and students. Some of the information is quite helpful; well-done sections targeting health professionals, women of color, and women with disabilities offer materials relating directly to specific needs and myriad links to reference tools. Under "Publications by Topic," however, many of the links are to standard government pamphlet materials on medical problems not unique to women. The "Frequently Asked Questions" (FAQ) section has some good responses to specific questions users have asked. The lively "What's New" section contains some full-text articles on women's health issues from popular magazines. The site also links to information on the agency itself.

Overall, this site is a supermarket offering consumers a vast array of information with varying degrees of value, in a variety of packages. Some of the information has traditionally been offered by government agencies; other kinds, such as FAQs, have only become possible via the interactiveness of the Web.

Another type of government site focuses on a specific national research initiative and contains links to each participating institution. An example is the National Centers of Excellence in Women's Health program <http://www. 4woman.org/owh/coe/index.htm>, another activity of the DHHS Office of Women's Health. It describes the initiative, links to every participating university hospital, and offers news releases and a "request for proposal" option. It is, in short, the public face of this program.

Teaching Hospital Megasites

Some teaching hospitals go to considerable effort to create megasites. Some of these provide extensive information to consumers, while others are geared more toward medical professions or students. All of them are good public relations, conveying the sense of a caring institution with great staff expertise. Mayo Clinic Health Oasis, an excellent site for health consumers, is produced by a staff of scientists, physicians, writers, and educators. Its "Women's Healthcenter section <http://www.mayohealth.org/mayo/common/ htm/womenpg.htm> offers information in several formats: "Headlines" are short articles about topics of current interest, such as liposuction and urinary tract infections. "Ask the Mayo Physician" is a FAQ. "Reference Articles" is a comprehensive collection of articles on topics such as vaginal dryness and weight training. A quiz section allows women to test their knowledge.

Links take the user to other sections: Cancer Center, Heart Center, Pregnancy and Child Health Center, and Diet/Nutrition Center. This site is especially impressive for its good editing. The design is clean; manageable amounts of information are presented (enough to be useful but not overwhelming); topics are interesting and timely; and the text is readable. It is updated weekly, and older materials are reviewed for currency (see also The Virtual Hospital, discussed below).

Nonprofit and Foundation Advocacy Sites

Sites in this category tend to advance the agenda of their sponsor. They are, predictably, narrower in scope than the sites discussed above. To varying degrees they reflect the philosophy of the sponsoring agency or institution and are not completely objective or neutral. However, if one takes into account the possible bias, these sites often provide high-quality information.

Women's Cancer Network <http://www.wcn.org/> is sponsored by the Gynecologic Cancer Foundation, an arm of the Society of Gynecologic Oncologists. Its motto is "Physicians dedicated to preventing, detecting, and conquering cancer in women." The site's aim is to help women "understand more about the diseases and treatment options, help them find appropriate cancer treatment specialists, and learn ways to prevent the development of cancer."[4] This goal is accomplished through various means: a directory where women can locate a gynecologic oncologist in their area; links for cancer survivors to the Web sites of support and advocacy groups such as Gilda's List; a questionnaire where women can identify their risk levels for breast, ovarian, endometrial, and cervical cancer; and, not surprisingly, a request for contributions to the foundation. This is a pleasant, well-designed site. Planned Parenthood (see below) is another example in this category.

Smaller Commercial Sites

Although the megasites of academic hospitals are ultimately commercial, it's easy to view them as a "public good"; their service orientation is highly developed. Some of the smaller commercial sites similarly provide high-quality information about their subject, above and beyond the goods or services they are selling. These, too, can be valid sources of information for consumers. One such site is sponsored by the Southern California Plastic Surgery Group <http://www.face-doctor.com> in Beverly Hills. It informs prospective clients in some detail about each of the cosmetic and reconstructive procedures offered: breast enlargement and reduction, "tummy tuck," facelift, and so forth. Photographs illustrate areas that would be removed or repositioned. Descriptions of procedures are objective and comprehensive. In addition, the site includes a section on what plastic surgeons do, introduces its medical staff, and provides directions and payment information. A variety of motives are doubtless at work here, including the sponsors' realization that an affluent and sophisticated clientele will feel positively inclined towards a medical group that helps inform them and presents itself via such a well-designed, up-to-date Web site.

Personal Sites

The Web's free environment encourages individual expression; some women have created highly personal Web pages about their own health issues. An example is Cancer Destroys, Cancer Builds, the thoughtful and intelligent Web site of Stephanie Byram, a doctoral student who was diagnosed with breast cancer <http://english-www.hss.cmu.edu/cultronix/stephanie/>. Her site is illustrated with memorable photos by Charlee Brodsky. One warning: personal sites, perhaps even more than other kinds of Web sites, can be ephemeral and may disappear without a trace.

WOMEN'S HEALTH, A SECTION OF HEALTHWEB

Background of the Women's Health Page

The remainder of this article will be devoted to a closer look at one specific women's health page. Women's Health <http://www.medsch.wisc.edu/chslib/hw/womens/womhome.htm> is part of HealthWeb, a collaborative project of the health sciences libraries of the Greater Midwest Region of the National Network of Libraries of Medicine, and those of the Committee for Institutional Cooperation. Currently, there are over twenty actively participating member libraries.[5] To ensure consistency among the 50 or so different sections, page authors follow a set of guidelines, and evaluate the Web sites they select, for quality. Completed pages undergo rigorous editing by a committee. The information on HealthWeb is intended for use by physicians, nurses, and other health personnel, as well as for health consumers.

In creating the Women's Health page in 1996-97, this author attempted to provide links to highly reliable Web sites–sources a user could trust in terms of the information provided or advice proffered. For each of these, a short description is provided, evaluating the link and informing users about who sponsors the site and what information they can expect to find there.[6] Finding reputable Web sites can sometimes be challenging. Women's health sites, to a greater extent than some other medical sites, run the gamut of credibility, from unimpeachable sources such as the Virtual Hospital to personal sites where women describe home remedies that have helped them. HealthWeb's Women's Health page starts with "General Resources," a list of megasites. Six of these are described below.

Centerwatch Clinical Trials Listing Service
<http://www.centerwatch.com/>

Centerwatch Clinical Trials Listing Service is a valuable resource for anyone needing information on clinical trials of new drugs. This would be a particularly good site for someone considering experimental chemotherapy. The site includes sections listing clinical trials currently recruiting patients, provides background on NIH clinical research studies in progress, offers a service that will notify you about new clinical trials, and features news updates on recent advances in clinical research. This site and Planned Parenthood (see below) exemplify the best kinds of Web pages mounted by organizations and agencies; they outline the agency's mission, provide a variety of information, and are frequently updated.

Global Health Network's Women's Health
<http://www.pitt.edu/HOME/GHNet/GHWomen.html>

Some sites on the General Resources page of Women's Health present collections of links. This site is an eclectic selection of links that are international in scope and also available in Japanese. A sampling of the topics covered includes: aging, domestic violence, nutrition and fitness, infectious diseases, and mental health. (The organization of this page served as a model for designing the Women's Health page.) Although still predominately American in origin, this site includes links from Canada, Mexico, Spain, Singapore, France, Germany, Sweden, and Japan. Some international conferences are also covered. These links offer Americans a chance to look beyond their own national concerns and observe how other nations are approaching their own health issues.

Michigan Electronic Library: Women's Health
<http://mel.lib.mi.us/health/health-women.html>

The Michigan Electronic Library is a virtual library designed for use by the libraries and residents of Michigan. It is selected and evaluated by a collaborative group of librarians at Michigan's state library system and the University of Michigan Library. The Michigan Electronic Library-Health Information Resource has several sections. "Evaluating Health Information on the Internet" links to a series of full-text articles that inform users how to better evaluate Web information. There is also an excellent list of links to major women's health sites. Some smaller sections on specific topics such as Childbirth, Breastfeeding, and Endometriosis offer several links apiece.

Planned Parenthood
<http://www.plannedparenthood.org>

Planned Parenthood is sponsored by the world's largest and oldest voluntary family planning organization. This is a complex (and somewhat visually confusing) site that offers many forms of information. A FAQ section provides answers to questions about pregnancy, birth control, and sexually-transmitted infections. Fact sheets cover such topics as teen pregnancy and abortion, and provide material from the Alan Guttmacher Institute (one topic is the donation of fetal tissue for treatment and research). An "Articles" section includes an essay entitled "What Most American Women Don't Know About the IUD." A "Sexual Health Glossary" offers short definitions of words such as Abstinence and Candida. This is a highly informative site that answers many, many questions.

Selected Non-Government Sources of Women's Health Information

Another broad-ranging list of links is the U.S. Food and Drug Administration's Selected Non-Government Sources of Women's Health Information <http://vm.cfsan.fda.gov/~dms/wh-ngov.html>. This site includes links to Web pages on topics ranging from breastfeeding to eating disorders to Chronic Fatigue Syndrome to sports injuries. Some of these sites are commercially sponsored, but all are serious and useful. This site illustrates the way the topic of Women's Health cuts across a broad range of other areas, bringing together a number of disparate yet related topics, some lying outside the traditional scope of medicine, such as domestic violence and fitness. This amorphousness may be, in fact, one reason the medical establishment may have been slow to recognize Women's Health as a separate field.

The Virtual Hospital

Some academic medical centers have created impressive sites filled with contributions written by their own specialists, some in the form of multimedia textbooks. These sites provide teaching tools for medical students and reference books for physicians, but they can also be valuable to consumers. How many of us have wished for access to information our physicians use? These sites tend to be rich with images (photographs, x-rays) and may include short videos illustrating the earmarks of a disease.

The Virtual Hospital <http://www.vh.org/>, one of the best of such sites, is sponsored by the University of Iowa's Hardin Library. To learn what cancers can spread to the breast, choose "For Healthcare Providers/ Information by Department/ Cancer Center" and open "Multimedia Textbook, Metastases to the Breast." If you're wondering if you might have the skin disorder Rosacea, look at a close-up photograph of a woman who does. Should you wish to read the "Clinical Practice Guidelines for Alzheimer's Disease," a document physicians use to check on current best-practice recommendations, choose "For Healthcare Providers/Information by Type: Clinical Practice Guidelines," and pick "Alzheimer's Disease." Virtual Hospital also contains a large section called "For Patients" containing less technical information. For example, if you were wondering how to interpret a recent pap smear report, you could go to the home page and select "For Patients/ Department/ Obstetrics and Gynecology" and open "What the Results of a Pap Smear May Mean."

WOMEN'S HEALTH SITES BY SUBJECT

The following section will discuss a sampling of the links available on the various "Specific Resources" subsections of HealthWeb's Women's Health Web site. Women's Health is *intentionally* a mid-sized site. It does not attempt to cover the full range of women's health issues comprehensively. Over time, new topics may be added. The topics covered are important ones, but may not include areas that some would consider essential. However, thousands of other Web sites are available, covering the entire spectrum of women's health issues.

Abortion

Abortion Clinics OnLine
<http://www.gynpages.com/>

Abortion Clinics OnLine is a directory of Web sites offered by over 200 providers of abortion services and other reproductive health care, both in the United States and internationally. The individual providers' Web sites provide details for patients. Clinics are also listed by the categories of abortion they perform. Also offered are articles about abortion rights, information on state parental consent laws, sources of financial assistance, and updated news.

The Association of Reproductive Health Professionals
<http://www.arhp.org>

This organization's multifaceted page is geared towards both professionals and consumers. For professionals, it offers notices of upcoming CME conferences and summaries of past conferences, as well as an international conference calendar. Legislative news (ARHP advocacy activity) appeals to both categories of users. For consumers, two interactive questionnaires gauge one's knowledge of contraception. A good listing of links provides information on breast, cervical and ovarian cancers; sexually transmitted diseases; pregnancy; infertility; abortion; menopause; men's health; and HIV-AIDS.

AIDS-HIV

HIV-AIDS Treatment Information Service
<http://www.hivatis.org>

The HIV-AIDS Treatment Information Service site is sponsored by a group of seven USPHS agencies. It furnishes updated treatment guidelines for adults and adolescents, children and newborns, as well as exposure guidelines for health care workers, and treatment guides for opportunistic infections and tuberculosis, both of which are frequently associated with AIDS.

Women and Safer Sex
<http://www.safersex.org/women>

Women and Safer Sex, one of the few Web sites offering safer sex information for lesbians, outlines sexual behaviors that are safe, probably risky, and very risky. The material on Women and HIV/AIDS contains statistics that were, as of this writing, quite dated.

Cancer

Benign Breast Disease and Breast Cancer Tutorial
<http://www.surgery.wisc.edu/wolberg>

This tutorial was designed by William H. Wolberg, MD for third-year medical students at the University of Wisconsin, but it is arranged so that consumers can alternatively select a listing of topics for lay people, if they prefer. Here you will find (among other information) informative sections on breast pain, the controversy over whether to recommend routine screening mammography for women in their 40s, and common reactions of children to the diagnosis of an adult's breast cancer, and how to help them.

Breast Cancer Resource Center
<http://www.healingwell.com/breastcancer/>

This versatile site has many features. The Online Information section covers such topics as the breasts, risk factors for breast cancer, side effects of treatment, and nutrition for cancer patients. Links to a selection of message boards enable users to share their experiences. Physician directories and a medical-insurance-plan directory are provided as well. A breast cancer medical news feature is available in Java format.

OncoLink: Gynecologic Oncology
<http://cancer.med.upenn.edu/specialty/gyn_onc/>

The OncoLink megasite is sponsored by the University of Pennsylvania Cancer Center and is one of the foremost cancer Web sites. This portion covers the following types of cancer: cervical and fallopian tube cancers, endometrial/uterine cancer, gestational trophoblastic disease and choriocarcinoma; ovarian, vaginal and vulvar cancers. Sections on each of the individual cancers share certain common features, but they differ to some degree in the other kinds of materials they offer. Generally speaking, there are links to support groups, physician information, patient information, links to clinical trials, question-and-answer sections, and abstracts of medical articles. One of the distinctive features of this site is the section "Confronting Cancer Through Art," a gallery of pictures created by people whose lives have been affected by cancer.

Reproductive Health

INCIID: International Council on Infertility Information Dissemination
<http://www.inciid.org/>

INCIID maintains this page, which is geared towards people with infertility problems who are seeking advice or support. "Medical Boards" (physicians) and "Support Boards" (peers) are forums where consumers can ask questions and comment on a broad range of infertility topics; chat rooms provide real-time support. Fact sheets written by professionals offer current information on a wide variety of topics, and there is a directory of fertility professionals.

Information about Contraception and Reproductive Health
<http://opr.princeton.edu/ec/contrac.html>

This site, sponsored by the Office of Population Research at Princeton University, is an extensive list of links to contraception and reproductive health Web sites. Categories include HIV/STD prevention sites (a number of these sponsored by government agencies), activist organizations, reproductive-health research organizations and journals, a source of films on human sexuality, and a condom page. Many of these sites are information-rich and geared towards lay users.

Domestic Violence

Domestic Violence
<http://www.en.com/users/allison/dvpage.html>

This site is a personal page by Allison Proctor. It defines domestic violence and links to several personal domestic violence pages. Proctor also discusses why and how people abuse, warning signs of abuse, a safety plan for leaving an abusive partner, and strategies for helping someone in an abusive relationship. Although a personal page, it is not out of place among larger, institutional pages because of its high quality and useful content.

Lesbian Health

Same Sex Domestic Violence
<http://www.xq.com/cuav/domviol.htm>

Same Sex Domestic Violence is sponsored by Community United Against Violence, a nonprofit agency focusing on violence directed against lesbian, gay, and transgendered persons. This site outlines the varying forms of abuse, offers advice to victims and friends of victims, and lists links to other Web resources.

Menopause

Doctor's Guide to Menopause Information and Resources
<http://www.pslgroup.com/MENOPAUSE.HTM>

This commercial site is sponsored by P\S\L Consulting Group, a Canadian medical communications and market research firm. It offers medical news and alerts such as a news release on the new Esclim estradiol patch; menopause information (what can I do about hot flashes and night sweats?); and links to discussion groups, newsgroups, and other menopause-related sites.

Osteoporosis

Clinical Trials: Osteoporosis
<http://www.centerwatch.com/studies/CAT111.HTM>

This site is offered by Centerwatch, a publisher focusing on the clinical trials industry. Information is provided on osteoporosis clinical drug trials currently open in the United States. Links are provided to research centers in this field, as well as to some good Web sites addressing musculoskeletal disorders.

Rheumatologic Disorders

National Institute of Arthritis and Musculoskeletal Diseases (NIAMS)
<http://www.nih.gov/niams>

This site is produced by NIAMS, a part of the National Institutes of Health. NIAMS heads up the federal research effort on the causes, treatment, and prevention of arthritis. Full-text brochures are offered on arthritis and exercise, arthritis pain, and other topics. Users are also given access to MED-LINE, information for prospective participants in NIAMS clinical studies at the NIH, and summaries of the valuable NIH Consensus Statements on optional calcium intake, total hip replacement, and sunlight/ultraviolet radiation and the skin.

Sexual Assault

Sexual Assault Information Page
<http://www.cs.utk.edu/~bartley/saInfoPage.html>

This high-quality personal page created by Chris Bartley presents a wealth of Web links on over thirty-five topics, including acquaintance rape, Rohypnol (the so-called date rape pill), incest, self-defense, abuse by professionals, and men's activism.

Sports and Fitness

Dr. Pribut's Running Injuries Page
<http://www.clark.net/pub/pribut/spsport.html>

Dr. Stephen M. Pribut is a specialist in Podiatric Medicine and Surgery. His Running Injuries page is designed for both men and women. A sampling of links includes: strength training for runners, side stitches, common running injuries, cold and hot weather running, etc. Some links are directed at women specifically; see, for example, "Women's Health and Fitness" and "Seeking Women's Health Resources on the Net."

CONCLUSION

The Internet offers consumers a colorful spectrum of sites in the relatively new field of Women's Health. Large teaching hospitals and governmental agencies sponsor megasites that present copious amounts of information in a variety of forms, some unique to the Web and others replicating paper media. Foundations and nonprofit-agency sites are narrower in scope and at their best, present useful, targeted material reflecting the values of the organization. Commercial sites can be valuable resources if they offer objective information. Personal Web sites provide fresh and unique perspectives on how individuals deal with medical problems. The Web site Women's Health, a section of HealthWeb, described in some detail above, offers one example of how the capabilities of the Internet can be mobilized to serve women's health needs.

NOTES

1. Boston Women's Health Book Collective. *Our Bodies, Ourselves: A Book by and for Women.* New York: Simon and Schuster, 1973.

2. Scott, Rebecca Lovell. "Women's Health." *NWSA Journal* 11(March 31, 1999):185.

3. Swenson, Norma. "A Life of Its Own–Looking Back at the Movement." *WomenWise: A Quarterly Publication of the Concord Feminist Health Center* 7(3):9.

4. Women's Cancer Network. Available: <http://www.wcn.org>.

5. This project is supported by the National Library of Medicine under contract #N01-LM-6-3523.

6. The HealthWeb Women's Health page will be undergoing periodic editing and updating under a new editor. This description represents its content as of January 2000.

Using Search Engines
to Locate Web Resources
on Women's Health

Elizabeth Connor

ABSTRACT. Despite the breadth, depth, and volume of information available on the Web, it can be tedious and time-consuming to locate and retrieve meaningful and reliable health content. This article describes selected Web sites devoted to the subject of women's health, mentions notable metasites developed by several medical libraries, showcases specific resources (answer services, search engines/directories, metasearch engines, and search voyeurs) for finding information, and discusses criteria for evaluating and critically examining Web sites. *[Article copies available for a fee from The Haworth Document Delivery Service: 1-800-342-9678. E-mail address: <getinfo@haworthpressinc.com> Website: <http://www.HaworthPress.com>]*

KEYWORDS. Women's health, search engines, World Wide Web, Web site evaluation

INTRODUCTION

Of the approximately 800 million pages comprising the publicly accessible World Wide Web, it has been estimated that only 16% have been indexed by search engines, and just 6% feature content related to education or sci-

Elizabeth Connor, MLS (connor@musc.edu), is Assistant Director of Libraries for Public Services and Education, Library, Suite 419, P.O. Box 250403, Medical University of South Carolina, Charleston, SC 29425.

[Haworth co-indexing entry note]: "Using Search Engines to Locate Web Resources on Women's Health." Connor, Elizabeth. Co-published simultaneously in *Health Care on the Internet* (The Haworth Press, Inc.) Vol. 4, No. 2/3, 2000, pp. 47-63; and: *Women's Health on the Internet* (ed: M. Sandra Wood, and Janet M. Coggan) The Haworth Press, Inc., 2000, pp. 47-63. Single or multiple copies of this article are available for a fee from The Haworth Document Delivery Service [1-800-342-9678, 9:00 a.m. - 5:00 p.m. (EST). E-mail address: getinfo@haworthpressinc.com].

ence.[1] Due to the sheer volume of Web-based resources, it can be tedious and time-consuming to locate and retrieve meaningful and reliable health content. This article provides brief descriptions of selected Web sites devoted to women's health, showcases specific search engines and directories, and discusses criteria for evaluating Web sites.

WOMEN'S HEALTH BACKGROUND

Although American women's life expectancies have exceeded men's since the turn of the century, due in part to decreased childbirth deaths, until recently women's health focused primarily on childbearing, menstruation, and menopause. American men suffer from more life-threatening chronic diseases, while American women suffer from acute and chronic diseases throughout their longer lifespans. Before recent efforts such as the U.S. Public Health Service's Task Force on Women's Health and the Women's Health Initiative <http://www.nhlbi.nih.gov/whi/index.html/>, women were routinely excluded from major clinical trials because of concerns about reproductive risks, even though some diseases are more prevalent or serious in women, and not all research data can be extrapolated from males to females.[2]

With increased focus and funding on diseases and conditions faced by women, and heightened interest in health information for women and men alike, knowledge of authoritative sites and effective use of search engines can save time and help improve understanding of diseases, conditions, and treatment outcomes.

EXAMPLES OF WOMEN'S HEALTH SITES

The following list is a representative sample of the rich variety of Internet resources related to women's health. On occasion, interesting sites can be discovered by word of mouth, direct advertising, articles such as this one, or by serendipitous means, while searching unrelated or tangential topics.

About.com–Women's Health
<http://womenshealth.about.com/>

Formerly known as the miningco.com, this site uses scores of human guides to select and annotate subject areas from art to travel. A freelance health writer who has compiled more than 100 topic areas from anemia to yeast infections maintains the guide for women's health. Highlights include health headlines, women's health forum, daily chat room, weekly newsletter, and a shopping area for related books and videos.

AllHealth–Women's Health
<http://www.allhealth.com/>

AllHealth (formerly known as BetterHealth) is a part of iVillage.com, a site developed for women (see Figure 1). Organized into sections such as conditions, wellness, diet, and experts, this channel's highlights include a drug database, MEDLINE, quizzes (stress, burnout, eating disorders), weight calculators, message boards, chats, and a virtual check-up. The left-hand frame search box allows searching of this channel, all of iVillage, or the snap.com site.

Discovery Health–Her Health
<http://discoveryhealth.com/>

Discovery Health, the consumer health site developed by the Discovery Channel television network, features a site called Her Health. Topics include women and heart disease, Viagra for women, and hormone replacement therapy. Her Health can also be searched with an internal search engine. Special features include a medical dictionary, and access to MEDLINE and other

FIGURE 1. AllHealth–Women's Health
<http://www.allhealth.com>

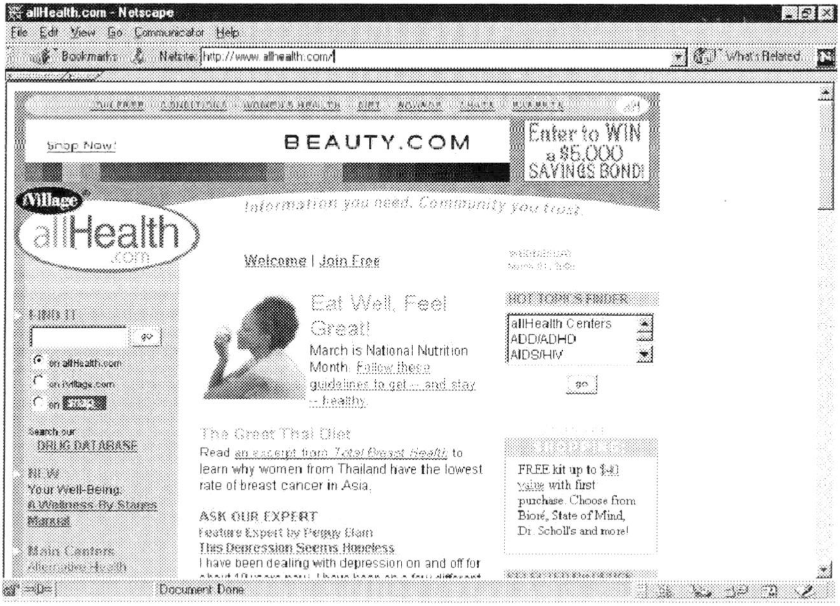

databases. The "Ask the Doc" section and MedCite, a topic-driven medical literature service, were developed by InteliHealth <http://www.intelihealth.com/>, a commercial enterprise from Johns Hopkins.

HealthAtoZ–Women's Health
<http://www.healthatoz.com/>

Special features include health headlines, research, and treatment updates (cancer, stroke, eating disorders), a message board devoted to women's health issues, various health tips (choosing a gynecologist, skin care, Kegel exercises), and a health e-card sending service. The interface can be individualized. E-mate is a portable health calendar and organizer. A search bar on the bottom of the page allows a focused search of more than 50,000 health-related Internet resources, reviewed by HealthAtoZ's Medical Advisory Board. Site experts include fitness trainers, pediatricians, nurses and librarians. Health AtoZ's sister site is MedConnect <http://www.medconnect.com/>, a resource for health professionals.

HealthGate
<http://bewell.healthgate.com/womenshealth/>

HealthGate includes a section for women only. Features include health news, medical tests, health calculators, a medical dictionary, and a health library with information about exercise, growing older, lifestyle improvement, treatment and medication, and other topics. The side frame allows basic and advanced searching of the site. MEDLINE, HealthSTAR, BIO-ETHICSLINE, and MDX Health Digest are among the databases that can be searched without charge.

JAMA Women's Health Information Center
<http://www.ama-assn.org/special/womh/>

Although the American Medical Association (AMA) developed this site primarily for physicians, consumers will be interested in top news stories, special reports, and a service called Women's Health Journal Scan with full-text content from AMA journals, and structured abstracts from non-AMA journals. *JAMA* uses the Excite search engine <http://www.excite.com/> to search for content within their information centers. The site is supported by Ortho-McNeil Pharmaceutical.

Mayo Clinic Women's Health Center
<http://www.mayohealth.org/mayo/common/htm/womenpg.htm/>

Developed by the renowned Minnesota-based clinic with facilities located in Jacksonville and Scottsdale, this site provides access to Mayo Clinic experts, health news headlines, full-text articles, quizzes, and other relevant content. The site search feature retrieves Mayo Clinic Health Oasis sites, recipes, and information from the U.S. Pharmacopeia Drug Guide. Other related Mayo health centers feature information about cancer, heart, pregnancy, and nutrition.

New York Times
<http://www.nytimes.com/specials/women/whome/>

The venerable *New York Times* offers women's health topics ranging from aging to urinary tract infections, aided by effective keyword searching of the site (see Figure 2). Highlights include statistics from the Centers for Disease Control and Prevention, an annotated guide to more than 100 women's health sites, news headlines, and a link to DiscoveryHealth <http://discoveryhealth.com>, which is described above.

FIGURE 2. *New York Times* Women's Health
<http://www.nytimes.com/specials/women/whome/>

National Women's Health Information Center
<http://www.4woman.gov/>

The National Women's Health Information Center is a project of the Office on Women's Health in the U.S. Department of Health and Human Services, and features content in both English and Spanish. The keyword search feature locates publications or organizations related to women's health. Sections include news headlines from Reuters, health-related legislation, and a health events calendar with upcoming conferences and special dates (February's National Eating Disorders Awareness Week, for example), of interest and importance to women.

OnHealth–Women's Health
<http://onhealth.com/ch1/resource/conditions/>

OnHealth, a Seattle-based consumer health information company, developed a women's health subsection within their site's "Conditions AtoZ" with information about more than 200 conditions from AIDS to yeast infections. Search capabilities include basic and advanced searching of selected sites, and free MEDLINE searching. OnHealth's Wellness Manager lets users customize their interface to include local weather, favorite discussion boards, and other content. Interactive tools include calculators (calorie, body mass index, sleep), tests (colon cancer risk, depression), an illustrated breast exam, and diet-specific food pyramids. Discussion subjects range from aches and pains, to weight gain/loss.

WebMD Health
<http://my.webmd.com/>

Although this site does not have a specific section for women's health, WebMD (see Figure 3) has attracted a great deal of attention among physicians, nurses, office managers, and consumers, touting itself as an "authoritative daily news source for health and medicine." WebMD's health channel hosts online communities; sponsors live Web events (such as jockey Willie Shoemaker talking about recovering from paralysis); and provides original news content in addition to reporting health news from Reuters and Associated Press. Please note: "WebMD may display search results based on monetary incentives provided by advertisers."

Yahoo–Women's Health
<http://dir.yahoo.com/Health/Women's Health/>

This site features nearly 30 categories from AIDS/HIV to female pattern hair loss in its Women's Health channel. From this page it is possible to

FIGURE 3. WebMD Health
<http://my.webmd.com>

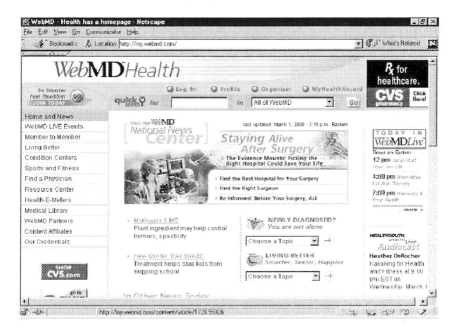

restrict results within Yahoo's women's health category, or search all of Yahoo. Yahoo uses the Inktomi search engine <http://www.inktomi.com/> to find resources outside its scope, as discussed below. Yahoo Health broadcasts (Dr. Laura Schlessinger talking about mental health is a typical example) are provided courtesy of WebMD <http://webmd.com>, which is discussed in greater detail above.

WEB SITES SELECTED AND ANNOTATED BY MEDICAL LIBRARIES

Over the past few years, academic health sciences libraries have selected and annotated links related to a wide range of subjects including women's health. Notable examples include Yale University's Cushing/Whitney Medical Library's Selected Internet Resources (SIR) <http://www.med.yale.edu/library/sir/> and the Hardin Library's MetaDirectory <http://www.lib.uiowa.edu/hardin/md/obgyn.html> at the University of Iowa.

HealthWeb <http://healthweb.org/> and NetWellness <http://www.netwellness.org/> are two notable academic collaborative efforts. HealthWeb is a collaboration of medical libraries of the Committee on Institutional Cooperation. Since first conceptualized in 1994, more than 20 midwestern libraries have shared responsibilities for selecting and organizing health-related Internet resources accessible through a common search interface. HealthWeb: Women's Health <http://www.medsch.wisc.edu/chslib/hw/womens> is maintained by librarians at the University of Wisconsin at Madison. It features general resources about women's health and selected resources about specific health topics pertaining to women. NetWellness <http://www.netwellness.org>, a joint effort from the University of Cincinnati, Case Western Reserve University, and Ohio State University, provides expert advice, health news, health topics from aging to violence, electronic databases, and full-text content (some of which is restricted to authorized Ohio users).

SEARCH ENGINES BACKGROUND

Although the Internet was developed in 1969 by the U.S. Defense Department <http://www.defenselink.mil/>, the World Wide Web was not developed until 1990. Gopher, the first Internet search tool, was developed in 1991, followed by the 1994 introduction of Yahoo <http://www.yahoo.com/>, the first Web search directory.

Search engines and directories vary according to database size, indexing practices, update frequency, and search syntax. Bigger is not always better, and for the purposes of seeking information on various health topics, directories may deliver more focused search results, especially for inexperienced searchers.

How do directories differ from engines? Technically, Yahoo <http://www.yahoo.com/> and About.com <http://www.about.com> are examples of directories, *not* engines, although it is possible to search the contents of each. Some directories use external search engines to find links outside the directory's catalog. Directories are created by hand, with human beings categorizing, ranking and/or annotating resources. Think of a Web directory as a *Reader's Digest* of sorts, filtering and organizing information for you. Directories are useful for browsing broad subject categories such as news, government, health and so forth. Engines can deliver good results when supplied with specific, meaningful keywords.

Some search engines have added directories to their search features, further blurring distinctions among them. Search engines use mechanical spiders, worms, or crawlers to visit Web sites and collect bits of information from the content and coding of sites. Search engines allow keyword searching of the entire World Wide Web, plus other resources such as discussion groups, media files, or news.

EXAMPLES OF SEARCH SITES

Answer Services

Several types of resources are available to retrieve and/or browse health information: answer services, and search engines/directories. Allexperts, Ask Jeeves, and LookSmart are examples of answer services.

Allexperts
<http://www.allexperts.com/>

Allexperts are volunteers with expertise in 1500 topic areas as diverse as car repair, health, travel, and music. While a number of physicians and other health professionals have volunteered their services to answer questions posed to the site, on topics as varied as abortion, epilepsy, and vascular surgery, some volunteers provide advice without professional credentials and may base their answers on personal experience and anecdotes.

Ask Jeeves
<http://www.askjeeves.com/>
<http://www.ask.com/>

Ask Jeeves (see Figure 4) allows natural language searching, meaning that it interprets and processes questions expressed in plain English. When Jeeves doesn't understand the query, it prompts you to express yourself differently or check your spelling. Typical queries include *where can I find information on lupus* or *drugs for ulcerative colitis*. First introduced in 1997, this resource tries to match questions with "answers" within its own database, and uses external search engines when it fails to make a match.

LookSmart
<http://www.looksmart.com/>

In addition to allowing natural language searching and searches posed as questions, LookSmart offers glimpses of what other users are searching (see more examples of voyeur engines below). LookSmart's personal health channel features live search editors that promise search help responses within twenty-four hours. Other categories include "Diseases/conditions," "Diet & Nutrition," "Drugs & Medicines," "Natural Therapies," and more.

Search Engines/Directories

With more than 1500 search engines in current existence, it is difficult to learn the features or capabilities of each. To stay aware of changes in size,

FIGURE 4. AskJeeves
<http://www.askjeeves.com>

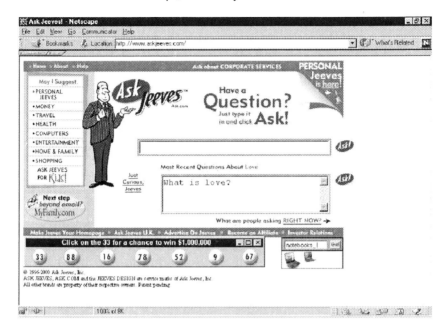

scope, or comparisons, consult Search Engine Watch <http://searchenginewatch. com/>. The following list of five engines/directories (AltaVista, FAST Search, GO/Infoseek, Google, Northern Light, and Yahoo) vary in size, complexity of interface, and available search options, and represent some of the most powerful tools available.

AltaVista
<http://www.altavista.com/>

First launched in 1995, AltaVista is one of the largest search engines, with many syntax features useful to novice and expert searchers alike, such as plus and minus signs, Boolean operators, parentheses to combine multiple ideas, and field searching (title, link, uniform resource locator). "Ask AltaVista" allows natural language searching; its results are derived from Ask Jeeves <http://www.askjeeves.com/> (see above). Estimated size: 250 million pages.

FAST Search
<http://www.alltheweb.com/>

Developed in Norway and launched in 1999, this engine was formerly known as All The Web. True to its name, this resource distinguishes itself with its sheer speed and simple search interface page. Search choices include all the words, any of the words, and the exact phrase. Estimated size: 200 million pages.

GO/Infoseek
<http://www.go.com/>
<http://www.infoseek.com/>

Jointly produced by Infoseek and Disney, GO is a portal site, offering superior search engine and site directory capabilities, and interface customization. Syntax features include using the pipe (|) to search within, plus/minus signs, quotations for searching phrases, and field searching (title, link, uniform resource locator, site). GO was launched in 1999, although Infoseek has been in existence since 1995. Estimated size: 75 million pages.

Google
<http://www.google.com/>

Google differs from other search engines in its use of link popularity to rank Web sites. If many site developers link to a specific health resource, for example, this may be considered a vote of confidence for the site's usefulness or unique qualities. Google offers a very simple, easy-to-use interface. Estimated size: 85 million pages.

Northern Light
<http://www.northernlight.com/>

Currently one of the largest Web indexes, Northern Light supports natural language searching, although use of specific keywords is more important than phrasing questions. This resource searches the Web and its own fee-based special collection. Special collection documents can be searched free, but viewing incurs charges per document. Interesting features include special limits (dates, subjects, or types of sites), field searching (title, uniform resource locator, and text within a site), and sorting by relevance and/or date. Estimated size: 189 million pages.

Yahoo
<http://www.yahoo.com/>

Although minuscule compared to other search resources, Yahoo more than makes up for its size with its easy-to-use and well-organized approach to information. Yahoo supplements its small catalog with results supplied by the

Inktomi search engine <http://www.inktomi.com/>. When it was introduced in 1994, it served as the first engine/directory. Its health channel includes a women's health section <http://dir.yahoo.com/Health/Women's Health/> (described above). Estimated size: 1.2 million pages.

MEDICAL SEARCH ENGINES

Although the search resources described above are adequate to locate and retrieve health-related sites, search engines specific to medicine and health permit greater search precision, and reduce extraneous and irrelevant results. For example, using a general search engine, a simple search on mumps might yield unrelated hits about MUMPS, a computer programming language.

Achoo Healthcare Online
<http://www.achoo.com/main.asp/>

Achoo (see Figure 5) is a powerful medical search engine developed and maintained in Canada. Features include "Site of the Week," "Reference

FIGURE 5. Achoo Healthcare Online
<http://www.achoo.com/main.asp>

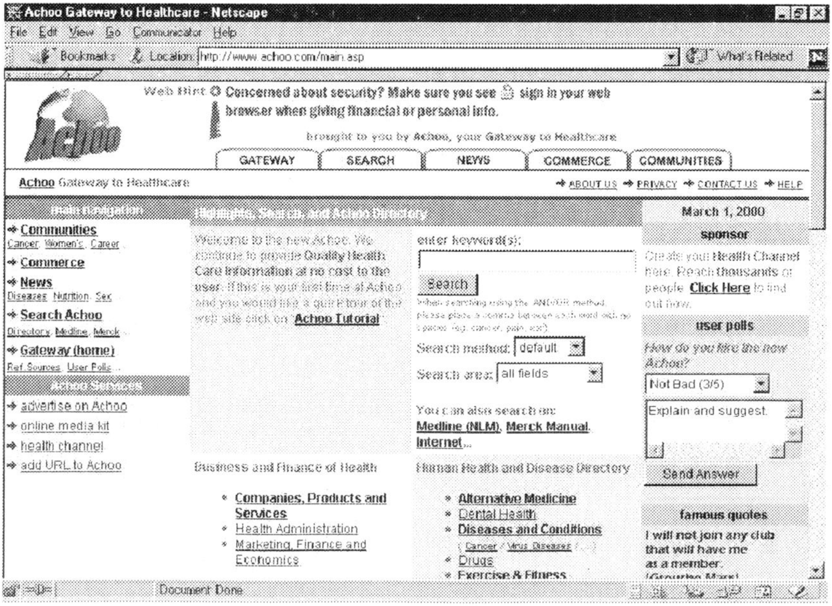

Sources" (journals, databases, directories, statistics), "Healthcare News," "Famous Quotes," voting polls and more.

CiteLine.com
<http://www.citeline.com/>

"Geared to professionals and serious consumers," this site is sponsored by Caredata.com, a health care market intelligence firm. Useful sections include limiting searches to "Disease & Treatment," "News & Journals," "Organizations," and "Research & Trials." The search engine is case sensitive.

MedExplorer
<http://www.medexplorer.com/>

Designed more for the health professional, MedExplorer (see Figure 6) includes internal search capabilities plus "Health-Exam" (drug database, MEDLINE, body mass index calculator, etc.), discussion forums and chat, more than 250 searchable health newsgroups, "Palm Health Zone" (hand-held computing applications), "Conferences," and more. Search categories range from alternative medicine to vision plus daily health tips. The "Ask the Doc" feature, also featured on Discovery Health <http://discoveryhealth.com/> is from InteliHealth <http://www.intelihealth.com>.

MedHunt
<http://www.hon.ch/MedHunt/>

Developed by the Health on the Net (HON) Foundation, this Swiss-based resource features content in both English and French. The database has two parts: HONoured sites that have been visited and annotated by humans, and the auto-indexed database generated by a mechanical spider. Capabilities include searching all the words, any of the words, and adjacent words. Displayed results can be limited to all sites, to HONoured sites, hospitals, support, and events, and by geographical region. The HON Code of Conduct <http://www.hon.ch/HONcode/Conduct.html/> is described below.

Medical World Search
<http://www.enigma.co.nz/mws/mws_source.htm/>

Considered by many as the first medical search engine, Medical World Search uses more than 500,000 medical terms from the National Library of Medicine's Unified Medical Language System to query other search engines

FIGURE 6. MedExplorer
<http://www.medexplorer.com>

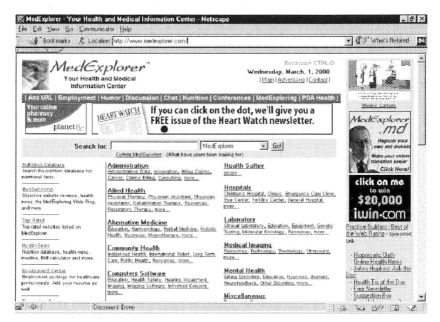

and to deliver focused results. Special features such as recalls of your last 10 queries or last 10 sites visited require a one-time user registration.

META-SEARCH ENGINES

Meta-search engines allow sequential or simultaneous searching of the same query in numerous search engines. The Big Hub <http://www.thebighub. com/> (see Figure 7), ProFusion <http://www.profusion.com/>, and SavvySearch <http://www.savvysearch.com/> are examples of effective sites that allow fast, simultaneous searching.

SEARCH VOYEURS

If you are curious about what or how other people search, try looking at one of the voyeur sites. Sites such as WebCrawler Search Voyeur <http://www.webcrawler.com/voyeur_wc>, Wordtracker <http://www.wordtracker.

FIGURE 7. The Big Hub
<http://www.thebighub.com>

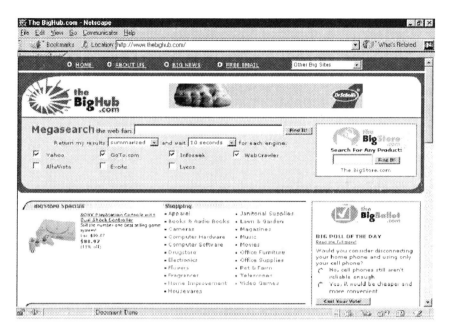

com/> and LookSmart Live <http://www.looksmart.com/live/> offer real-time search glimpses, similar to a stock market ticker activity. Other sites such as MetaCrawler MetaSpy <http://www.metaspy.com/> and Ask Jeeves Peek Through the Keyhole <http://www.askjeeves.com/dosc/peek/> show samples of recent search terms, updated every 15 to 30 seconds. For top search queries, compiled each week or month, visit Lycos Top 50 <http://50.lycos.com/> or SearchTerms <http://www.searchterms.com/>.

USING SEARCH ENGINES TO FIND INFORMATION ON WOMEN'S HEALTH

Let's consider how an understanding of Boolean logic <http://www.newsbank.com/whatsnew/boolean/> can help us craft better searches and improve our understanding of how search engines work. Simply said, AND is used to combine two or more concepts. For example, a search of *women AND lupus AND treatment* results in documents that include all three concepts. AND *decreases* or narrows retrieval. In some search engines, the plus sign (+) achieves similar results, for example, *+women +lupus +treatment*.

OR is used for either concept. For example, *chickenpox OR mumps* retrieves material on either subject. OR *increases* or broadens retrieval. In some search engines, the pipe symbol (|) or colon (:) achieves similar results. For example, *chickenpox | mumps* or *chickenpox:mumps.*

NOT eliminates a concept. For example, *vaccination NOT polio* finds information about other immunizations. NOT *decreases* or narrows retrieval. In some search engines, the minus sign ($-$) achieves similar results, for example, *vaccination-polio.* Use this capability with caution as sites covering broad subject areas could be unintentionally eliminated.

Librarians and other information scientists use advanced search features to achieve large search results, and in some instances, more is better. Skilled searchers retrieve large sets, combine them mathematically, and filter material down to a smaller set of relevant results. Over time, search engines will get better at focusing and filtering information.

For typical Web users, however, numerous hits can prove disastrous, especially if relevancy is more important than comprehensiveness, and if time is a major consideration. The most common searching pitfalls relate to spelling and punctuation, understanding specific search interfaces and syntax rules, using too many specific keywords, or vague terms such as illness or disability. Search terms such as eating, diet, and nutrition can be considered vague, especially if one were seeking information about breakfast foods, for example.

CONCLUSION

Frequent visitors to medically related sites might notice a designation from the Health on the Net (HON) Foundation. In addition to providing an English/ French medical search engine <http://www.hon.ch/MedHunt/>, this Swiss-based organization uses various criteria to "honour" sites that conform to the HON Code of Conduct. This code <http://www.hon.ch/HONcode/ Conduct. html> relates to revealing qualifications for providing medical advice, maintaining confidentiality, referencing source data, indicating the date information was last modified, providing contact information, identifying funding or content sources, describing the site's advertising policy, and differentiating between promotional content and original content.

Other criteria can factor into the relative usefulness[3] and authoritativeness[4] of resources available on the Internet and used by consumers to better understand health and disease:

- *What is the authority of the content creator?* Is the creator a lay person with the disease or a world expert in the diagnosis/treatment of the same? For what audience is the material intended? Casual bystander, student or researcher? Current or future customer? Third grader or university professor?

- *Accuracy of the information* (facts, spelling, current practice) goes without saying, but Internet searchers are bound to find outlandish claims, unsubstantiated or undocumented information, falsehoods, gross generalizations, and harmful information among the good and useful.
- *Currency of the information.* When was the content posted? When was the last site update?
- *Consider the point of view or bias (if any) of the content developer.* Who developed the content? Is the author the same as the publisher? Does it matter? Can you contact the author for more information?
- *Did the publishers obtain permission to publish this information on the Internet?* Does the site have a disclaimer or copyright statement?

In conclusion, although many good sites abound related to women's health, many more can be discovered by learning to use one or two good search engines. The challenges for the next few years will be to examine and transform how consumers seek and use essential health information, possible now through the use of search voyeur sites and other means. Interfaces can be streamlined to reduce barriers to access and eliminate unnecessary search steps. More work needs to be done to add value to the information retrieved by developing innovative and collaborative resources such as HealthWeb <http://healthweb.org/> or NetWellness <http://www.netwell.org/>.

NOTES

1. Lawrence, Steve, and Giles, C. Lee. "Accessibility of Information on the Web." *Nature* 400(July 8, 1999):107-9.

2. Bird, Chloe E., and Rieker, Patricia P. "Gender Matters: An Integrated Model for Understanding Men's and Women's Health." *Social Science and Medicine* 48(March 1999):745-55.

3. Rettig, James, and LaGuardia, Cheryl. "Beyond 'Beyond Cool' Reviewing Web Resources." *Online* 23(1999):51-5.

4. Kim, Paul; Eng, Thomas R.; Deering, Mary Jo; and Maxfield, Andrew. "Published Criteria for Evaluating Health Related Web Sites: A Review." *BMJ* 318(March 6, 1999):647-9.

Link by Link:
Building NOAH's Women's Health Page

Jane Fisher

ABSTRACT. New York Online Access to Health (NOAH) is an authoritative bilingual health information resource developed in 1995 by four New York City partners: The City University of New York, The Metropolitan Library Council, The New York Academy of Medicine, and The New York Public Library. NOAH's mission is to provide high-quality full-text information for consumers that is accurate, timely, relevant, and unbiased. This paper discusses NOAH's Women's Health page, how it is arranged for easy access, the selection process for adding content, and the criteria used to evaluate documents and links. NOAH can be found at <www.noah.cuny.edu>. *[Article copies available for a fee from The Haworth Document Delivery Service: 1-800-342-9678. E-mail address: <getinfo@haworthpressinc.com> Website: <http://www.HaworthPress. com>]*

KEYWORDS. NOAH, New York Online Access to Health, women's health, Internet

INTRODUCTION

New York Online Access to Health (NOAH), an easy-to-navigate full-text health information resource, began in 1995 when the U.S. Department of

Jane Fisher, MLIS (jfisher@nypl.org), coordinates "CHOICES in Health Information," The New York Public Library's consumer health information service, 455 Fifth Avenue, New York, NY 10016. Jane presently serves as the Secretary of the Consumer and Patient Health Information Section (CAPHIS) of the Medical Library Association. She is the Contributing Editor of NOAH's AIDS/HIV page and the Women's Health Page.

[Haworth co-indexing entry note]: "Link by Link: Building NOAH's Women's Health Page." Fisher, Jane. Co-published simultaneously in *Health Care on the Internet* (The Haworth Press, Inc.) Vol. 4, No. 2/3, 2000, pp. 65-74; and: *Women's Health on the Internet* (ed: M. Sandra Wood, and Janet M. Coggan) The Haworth Press, Inc., 2000, pp. 65-74. Single or multiple copies of this article are available for a fee from The Haworth Document Delivery Service [1-800-342-9678, 9:00 a.m. - 5:00 p.m. (EST). E-mail address: getinfo@haworthpressinc.com].

Commerce awarded a National Telecommunications and Information Administration grant to NOAH's original partners, The City University of New York, The Metropolitan Library Council, The New York Academy of Medicine, and The New York Public Library. The grant's task was to develop an authoritative bilingual health information site using library venues that provided access to health information for the underserved. Since that time, NOAH has received numerous awards and logs over 500,000 visits per month.[1]

NOAH's mission, simply stated, is to provide high-quality full-text information for consumers that is accurate, timely, relevant, and unbiased. In the 1998 edition of *Consumer Health Information Source Book,* Alan Rees gave NOAH an outstanding ranking and said that NOAH is "a vast treasure trove of reliable consumer health information on AIDS, Alzheimer's disease, asthma, cancer, diabetes, heart disease and stroke, STDs, tuberculosis, and much more. Bilingual English-Spanish. First rate!"[2] There are currently more than forty different NOAH Health Topics pages, including Aging and Alzheimer's Disease, AIDS, Alternative Medicine, Asthma, Cancer, Dentistry, Environmental Health, Eye/Vision, Headaches, Heart Disease, Lyme Disease, Neurological Disorders, Pregnancy, Women's, Men's and Children's Health, and many others (see Figure 1). The content provided on NOAH's Health Topics pages is mined from a variety of respected and authoritative sources. This paper discusses NOAH's Women's Health page, which is accompanied by pages for Men's Health and Child and Teen Health under the heading "Personal Health."

Librarians and medical information specialists participate in the development of NOAH by volunteering to be contributing editors. Contributing editors assume the responsibility for organizing and maintaining a page on a specified topic. Based on a set of selection criteria, the contributing editors examine full-text documents for potential inclusion in the NOAH topics page for which they are the assigned editor. In November 1999, NOAH's Women's Health page was expanded to include additional topics and a greater variety of full-text documents (see Figure 2). The page arrangement, selection process, and evaluation criteria used to build the Women's Health page are described below.

PAGE ARRANGEMENT

All NOAH pages strive to provide clear and simple navigation, the goal being that with only a few mouse clicks, users find the desired information. The Women's Health page is arranged in an outline format with the major headings "Basics," "Specific Issues," and "Information Sources." In the "Basics" section, consumers find easy-to-understand guides to the female

FIGURE 1. NOAH Health Topics Page

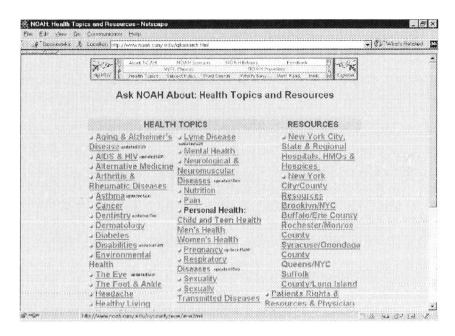

reproduate system and a glossary to puberty and menstrual health. The "Specific Issues" section is comprised of an alphabetized list of women's health concerns and includes information about breast disorders, cervical conditions, hysterectomy, lupus, menopause, menstruation, osteoporosis, toxic shock syndrome, sexually transmitted diseases, uterine disorders, vaginal disorders, and vulvodynia. When a related topic has its own NOAH page (e.g., Pregnancy, Birth Control, and Cancer are separate NOAH Pages), users are pointed directly to those NOAH pages from the Women's Health page. "Information Sources" is the final portion of the Women's Health page. This includes links to general information and news sources such as the *New York Times* Women's Health section, the National Women's Health Information Center, The National Women's Health Resource Center, and the Office of Violence Against Women. Links to related voluntary organizations are also provided in the "Information Sources" section.

To assure that NOAH is easily accessible to the Internet community at large, its navigation remains purposefully simple and avoids the use of frames and slow-loading graphics. Portable Document Format (PDF) files, however, are included, and information about how to download the Adobe Acrobat® Reader to view PDF files is readily provided. Sources that charge a

FIGURE 2. NOAH Women's Health Page

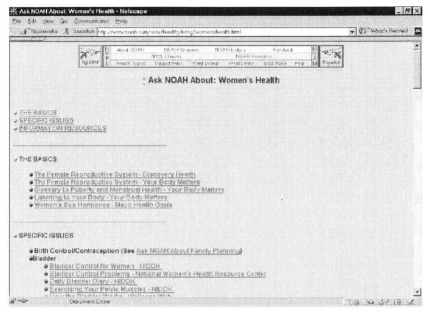

fee for use are generally omitted, especially if the content can be easily found on free sites. An Excite® word search feature is provided by NOAH to help users who do not find the information they need from a quick scan of the outline, which serves as a table of contents. For example, if a user is looking for information about a pap smear but does not suspect that this information will be located under the cervix heading, a word search will point the user in the right direction. Because NOAH emphasizes clear and simple navigation and organization, word searching is optional and the desired information is usually found directly from the outline with only a limited number of mouse clicks. In addition, "see" and "see also" reference links are used throughout NOAH pages to assure that users find all relevant documents.

SELECTING CONTENT

Selecting content for inclusion in each NOAH topics page is the most important responsibility of NOAH's contributing editors. Contributing editors cull through what has become the morass of consumer health information sites in search of the best information on each specific aspect or condition covered by the page. NOAH's goal is to provide full-text documents that address consumer health questions, not mere lists of links. Contributing editors use the knowledge they have from working at library information desks to anticipate NOAH users' information needs. In general, the rule of thumb "less is more" applies to NOAH pages. Editors make no attempt to include every document about the specific issues; instead they seek out sites that offer unique information about the specific issues or concerns.

Each full-text document considered for inclusion on a NOAH page is examined by the NOAH contributing editor to assure that it meets a series of selection criteria. A blind peer review process for each new NOAH page is also required before any new topic is added or any existing page is substantially revised. The criteria used to evaluate documents for inclusion in NOAH are authority, accuracy, currency, bias, and readability. They are described below and can easily be used to evaluate other women's health information Web sites.

Authority

Contributing editors look to see if a specific author is cited and whether that author is considered to be an expert in the field. If no individual author is cited, editors will examine the document for other identifiable sources or affiliations. An "About this Organization" link or an opportunity to "Contact the Webmaster" can help verify a site's authority. Documents that fail to provide a reliable individual author or affiliation are generally omitted by NOAH editors. Sources for NOAH's Women's Health page were initially identified by scanning the Web sites of reputable voluntary agencies, government offices, foundations, and online health books. Examples include the American Academy of Family Physicians, the National Institutes of Health, the American Women's Health Association, the American Cancer Society, the National Women's Health Information Center, and the Home Edition of the Merck Manual. The author or affiliation of documents from these types of sources is readily identified and verified. These sites can also act as a springboard for identifying other sources of reliable information. When the author or affiliation is identified, the question then becomes whether the author or agency has the appropriate background or expertise to advise on a particular subject.

Accuracy

Even with a known author or affiliation, it is important that all documents be reviewed for accuracy. Pages with errors or questionable links, promises of quick cures, and even pages with excessive or distracting advertising are generally passed by for something that better meets NOAH's mission. Because the accuracy of the information is so important, NOAH does not include links to chat rooms and newsgroups whose content changes rapidly and cannot be verified. Opportunities to communicate with others in an open forum that online conversations offer undoubtedly give some Internet users solace and comfort. However, because the content of these forums is not monitored for accuracy, the risk of perpetuating false information is great. Information about support groups by geographic area and contact information for online mailing lists, however, is available on NOAH.

Currency

New advances and constant changes in medicine make it essential that information about health, whether from online or print sources, be current. Since the Internet is unregulated and unmonitored, it is easy to stumble across Web sites that sit unattended and are out of date. NOAH users have come to rely on NOAH for its access to current information. Reviewing the document under consideration for inclusion on the NOAH Women's Health page for the date it was last updated is one quick way to determine whether it is being kept current by its author. A document's date is usually found in the footer at the bottom of the page. Documents that are undated or have not been updated in more than twelve months are dubious. Occasionally, however, the information is so basic that it is unlikely to change. In these cases, the document may be included even if it has not been recently updated.

NOAH contributing editors are responsible for keeping their assigned health topics pages up-to-date. Links are manually checked every six months. Broken links are repaired or removed, and any information that has become dated is replaced. In addition to checking for a date on a document that is being considered for inclusion on a NOAH page, NOAH's contributing editors also attempt to stay informed about their subjects so they can judge the currency and accuracy of a document. The editor of the Women's Health page, by way of example, should be knowledgeable about the current guidelines for mammography, the range of treatment options for uterine fibroids, and the latest schools of thought on hormone replacement therapy to assure that the most current sources of information are being included.

Bias

Any evidence of bias that can even potentially compromise the value of the information is suspect. Documents considered for NOAH pages are reviewed to determine if a particular political, social, or philosophical slant is present. Sometimes information that appears to be straightforward and unbiased at first glance can reveal a subtle advertising message upon closer inspection. For example, if a Web site about varicose veins is authored by a clinic that specializes in laser therapy, the site will be carefully reviewed to assure that it does not tout laser therapy as the only approved treatment option. In some instances, bias can be reduced by adding links to additional documents that speak from opposing points of view as a way of supporting a more balanced picture. Web sites that only serve to advertise specific products or treatments under the guise of providing consumer health information are generally omitted.

Contributing editors also watch out for potential bias in the agency that sponsors or hosts a Web site. If all the hyperlinks from a document point to a particular political point of view, or point to product sales information, a different source will be sought. Similarly, NOAH might omit a source whose sponsoring agency's mission conflicts with NOAH's desire to provide unbiased information to consumers. An online pamphlet about contraception that describes a wide range of birth control options, for example, will be preferred to a source that advocates abstinence as the exclusive method of birth control.

Readability

NOAH is a consumer health site and its audience is the general public, so the ability of the layperson to understand the content of a Web source is another important criteria for evaluating online information. Recent literature about illiteracy and health care reveals that 50% of the population may be unable to read and understand basic medical instructions.[3] Additionally, a review of the medical materials generally recommended for consumer health collections in public libraries showed that these materials far exceed the reading levels of the general public.[4] NOAH makes every attempt to include information that is written for the average adult reader and is currently seeking to expand its audience by providing information written specifically for teenagers and children. In some NOAH health topics pages, scholarly and research level information is included. NOAH's breast cancer page, for example, includes not only consumer level information, but also offers *PDQ Information for Health Care Professionals,* which is a research level online information service. When scholarly information is selected for NOAH pages, the link is labeled *Physicians* to alert users to the fact the material is indeed scholarly. Because NOAH is first and foremost a consumer health information product, links to research level information are only included as a supplement to consumer sources, never as a replacement.

WOMEN'S HEALTH EN ESPAÑOL

Since its launch in June of 1995, NOAH has established its presence on the Web as an outstanding bilingual consumer health information site. Visitors to NOAH can access reliable and authoritative Spanish language health information by simply clicking on the "Español" button available from all NOAH Topics pages on the top and bottom banner. The Spanish Women's Health page (see Figure 3) offers the same arrangement as the English version, with variation in the actual content of the documents available. Specific information topics provided in Spanish include cancer, conditions and infections, reproductive health, pregnancy, menstrual functions, nutrition, sexuality, and AIDS and HIV.

CONCLUSION:
PUTTING ONLINE HEALTH INFORMATION TO WORK

When NOAH was conceived in 1995, its anticipated audience was populations with low socioeconomic status who depended on public libraries for

FIGURE 3. La Salud de la Mujer

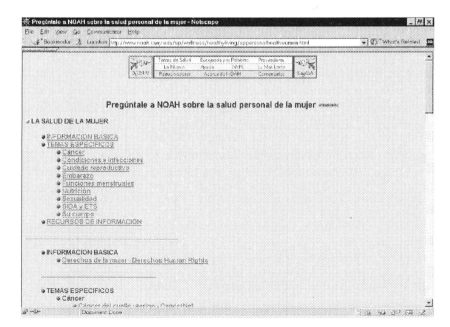

access to the Internet. As the Internet's popularity as a source for health information soars, NOAH is used by people around the world who have discovered that it is a quick and easy source for reliable and current information. Access to NOAH and other similar quality-filtered health sites can save users from the time and frustration of looking for health information by surfing the Web or using broad based commercial search engines. The Internet is one of the last unregulated frontiers in our culture and while this is good news for democracy, it can be bad news, even dangerous, for the neophyte Web surfer looking for reliable sources of online health information. Rees concurs, stating, "Surfing, haphazard almost by definition, is more suitable for embarking on a fascinating voyage of exploration than for searching for information in a systematic, rapid, and efficient manner."[2]

Health information provided by NOAH is not intended to replace the role of the health care provider, and NOAH makes no attempt to answer users' individual health questions. A disclaimer appears in the footer of NOAH pages stating that "NOAH is an information guide only and cannot answer personal health-related or research questions. NOAH's information has been culled from a variety of consumer health resources; it is offered to you with the understanding that it not be interpreted as medical or professional advice. All medical information needs to be carefully reviewed with your health care provider."[5] Women's health information found on the Internet, like any other health topic, is best used in conjunction with qualified medical care. Reliable and current documents found online can serve to enhance the quality of communication with health care providers and help consumers be active participants in their health care.

NOAH's current partners include The City University of New York, The New York Public Library, The Metropolitan New York Library Council (METRO), The New York Academy of Medicine, The Queens Borough Public Library, The March of Dimes/Birth Defects, Inc., Aetna US Health-Care, New York University Medical Center, St. Luke's Hospital-Roosevelt Hospital, and The Brooklyn Public Library. NOAH is a nonprofit site and is supported entirely through sponsorship programs and grants.

NOTES

1. The NOAH Team. Press Releases–NOAH Background. [Web document] New York: The Team. Available: <http://www.noah.cuny.edu/press/press.html#Background>.
2. Rees, A. Consumer Health Information Source Book. Phoenix, AZ: Oryx Press, 1998.

3. Kefalides, P.T. "Illiteracy: The Silent Barrier to Health Care." [Web document] *Annals of Internal Medicine* 130(February 16, 1999):333-6. Available: <http://www.acponline.org/journals/annal/16feb99/currilli.htm>.

4. Baker, L. et al. "Consumer Health Materials Recommended for Public Libraries Too Tough to Read?" *Public Libraries* 35(March/April 1996):124-30.

5. The NOAH Team. Disclaimer. [Web document] New York: The Team. Available: <http://www.noah.cuny.edu/qksearch.html>.

The Perils of Pauline:
Women as Health Care Consumers

Janet M. Schneider

ABSTRACT. Women face special health care challenges that are directly related to their gender and often result in health care that differs greatly from that of their male counterparts and adversely affects personal health and quality of life. This article discusses the skills women need to become active, responsible advocates for their own health: effective communication techniques, finding the qualifications of their doctors and hospitals, personal medical record-keeping, legal rights and responsibilities, insurance options, and how to find additional health information. *[Article copies available for a fee from The Haworth Document Delivery Service: 1-800-342-9678. E-mail address: <getinfo@haworthpressinc.com> Website: <http:// www.HaworthPress.com>]*

KEYWORDS. Women's health, consumer health, health care providers, patient advocacy

INTRODUCTION

Patients expect their health care providers to quickly diagnose illnesses, prescribe the right test/medicine/treatment, and solve all health problems. But

Janet M. Schneider, MA (janet.schneider@med.va.gov), is Patient Education Librarian, James A. Haley Veterans' Hospital, Tampa, FL. She has been active in patient health education programs and activities for the Department of Veterans' Affairs at the local, network, and national levels for 22 years, and is Senior Editor of the VA's Florida and Puerto Rico's Sunshine HealthCare Network *Sunshine Health-Net*, a quarterly health information newsletter for veterans. She has critically reviewed health information books for *Library Journal* since 1992.

[Haworth co-indexing entry note]: "The Perils of Pauline: Women as Health Care Consumers." Schneider, Janet M. Co-published simultaneously in *Health Care on the Internet* (The Haworth Press, Inc.) Vol. 4, No. 2/3, 2000, pp. 75-83; and: *Women's Health on the Internet* (ed: M. Sandra Wood, and Janet M. Coggan) The Haworth Press, Inc., 2000, pp. 75-83. Single or multiple copies of this article are available for a fee from The Haworth Document Delivery Service [1-800-342-9678, 9:00 a.m. - 5:00 p.m. (EST). E-mail address: getinfo@haworthpressinc.com].

the truth of today's health care, particularly for those insured by managed care plans, is that individuals must be proactive, informed health care consumers. The passive patient of yesterday can no longer expect health care providers to bestow good care without her active participation. This is evidenced in the National Academy of Science's Institute of Medicine recently released report,[1] stating that at least 44,000, and perhaps as many as 98,000, Americans die each year as a result of medical mistakes. These mistakes include medication errors, inappropriate treatments due to incomplete medical records, wrong-site surgeries, and mistaken patient identities. While some situations can only be corrected by health care providers, patients can learn skills that may prevent many of these types of errors through assertive communication and engagement in their own health care. The individual patient must use those skills to be an informed and wise "shopper" through the sophisticated maze of specialized medicine, to ensure her own quality health care.

According to the Beijing Platform for Action, adopted at the Fourth World Conference on Women in 1995, "women have the right to the enjoyment of the highest attainable standard of physical and mental health."[2] For a variety of reasons, however, women cannot expect to attain that standard without becoming educated about their health rights, responsibilities, and legal options–and acting upon that knowledge. While all individuals, regardless of sex, should learn the skills necessary to deal with health care providers and systems, women face special challenges directly related to their gender. These challenges often directly result in health care that differs greatly from that of their male counterparts, and adversely affects personal health and quality of life.

What are these challenges and are they really that different from those that men face in health care? There is, of course, the obvious distinction of the sex itself. Physicians have traditionally treated normal functions, such as pregnancy and menopause, as medical problems rather than natural phases of a woman's life. These functions, while normal, still cause considerable concern for many women, while not receiving appropriate attention from physicians. Many women don't even obtain adequate preventive screening at their annual gynecologic check-ups. Unless the physician asks about exercise and nutrition, and risk factors for STDs, cancer, heart disease, etc., women are not receiving full benefits for their time and money. Other risk factors include:

- The bias of health care providers and systems toward women. Most medical research has included only male subjects, ignoring females almost entirely. A patient's gender influences the physician's choice of diagnostic tools and treatment options, with men frequently being offered more options and aggressive treatments.[3] Women's reproductive options can also be limited by where they receive their care; Catholic-owned hospitals, for example, enforce their own rules for contraception, fertility treatments, and abortion.

- Women's financial and educational levels.[4] Access to health care often depends on having employment that offers health insurance, and good jobs require minimum levels of education. Poverty is a major cause of ill health and early death, simply because women cannot afford health care. Despite the strong economy, women still generally earn less than men, and more women are uninsured today than in 1993. Even with insurance, many women can't afford to take time away from their jobs for illness or preventive health check-ups.
- The levels of violence in women's lives. The Commonwealth Fund 1998 Survey of Women's Health found that two of every five women reported at least one experience of abuse or violence in their lifetimes.[4] Abused women have been shown to be at higher risk for psychological and physical health problems, necessitating additional uses of the health care system.
- Caregiving roles.[4] In 1998, nine percent of all American women were caring for a sick or disabled family member. The added responsibility of caregiving places enormous physical and emotional health strains on women who are already burdened with employment and child-rearing responsibilities.
- Increased contact with the health care system.[4] Women visit their health care providers more often than men, and make most family health care decisions. They also, as caretakers, become responsible for taking their children and other family members to doctors as needed.

Women must become active, responsible advocates for their own health. Since women have become the primary family negotiators through the health care system, it is critical that they learn

1. effective communication techniques;
2. how to find the qualifications of their doctors and hospitals;
3. personal medical record-keeping;
4. their legal rights and responsibilities, and insurance options; and
5. how to find additional health information.

While librarians and other information professionals can (and should) direct women in locating a host of print resources, women can also acquire a wealth of information from the Internet. A recent CommerceNet/Nielsen Media Research Study <http://www.commerce.net/news/press/061699.html> showed that of today's 92 million Internet users, 46% are women. Of this population, 64% work full time, and 54% are mothers. Many of these women find it difficult to discuss sexual issues, eating disorders, and other "private" topics with their health care providers. The Internet provides an anonymous medium for women to garner needed health information and support in the privacy of their own homes.[5] As more women acquire access to computers, the Internet provides a great educational opportunity to transform the "medically illiterate" into informed, active, and responsible health care consumers.

FINDING QUALIFIED HEALTH CARE PROVIDERS AND SYSTEMS

It is assumed that all health care providers are qualified and competent. But with the explosion of research and technologies, most physicians are unable to keep up with the enormous number of new treatments, guidelines, and protocols.[6] Physicians base a large number of treatment decisions on their own experience and original training, which may be limited and/or outdated. Unless they are willing to spend a considerable amount of time on continuing education, their knowledge and skills suffer. It is therefore important for patients to examine their health care providers' educational credentials for accreditation, board certification, date of degree, continuing educational status, and where he/she has hospital privileges. Patients must also acquire the tools to utilize for evaluating the quality of health care facilities and personnel.

An excellent place to start is at the Agency for Health Care Policy and Research (AHCPR) <http://www.ahcpr.gov>. The AHCPR is a branch of the Department of Health and Human Services and is charged with supporting research designed to improve the quality of health care. The information it provides is evidence-based and available to health care providers and consumers. Its 47-page booklet, "Your Guide to Choosing Quality Health Care" <http//www.ahcpr.gov/consumer/qntool.htm>, is available as a PDF or HTML file, and contains tips and worksheets designed to help the consumer choose health plans, doctors, treatments, hospitals, and long-term care facilities. The HTML version is designed for those who wish to use portions of the material in newsletters, brochures, or other print/electronic publications. JPG image files are included.

The AHCPR has other resources to assist consumers, from the pamphlet "Choosing Quality in Health Care," to links for clinical practice guidelines <http://www.ahcpr.gov/clinic/index.html>. The clinical practice guidelines are important for consumers who, after receiving a diagnosis, wish to compare their treatment with the evidence-based standards of the guidelines. Users may also print their own "Personal Health Guide," which provides information on prevention screenings and a fill-in-the-blank record for weight, cholesterol, blood pressure, and so on. Links are provided to MEDLINE*plus,* the National Women's Health Information Center, and healthfinder for researching specific health conditions. Other documents include information on types of health insurance (including special low-cost or free insurance for children), questions to ask prior to surgery, and common uterine conditions.

The Joint Commission on Accreditation of Healthcare Organizations (JCAHO) <http://www.jcaho.org/> enables users to "Search for Quality Information on Health Care Organizations." The search capabilities enable the user to distinguish how hospitals, health care networks, long-term care facilities, or laboratories meet the Joint Commission standards, which are stringent and evaluate all aspects of the facilities' operations. While this is not always the best litmus test for quality, facilities that fail the standards must be viewed–and used–with caution.

The American Medical Association's Consumer page <http://www.ama-assn.org/consumer> contains a "Doctor Finder" and a "Hospital Finder" as well as a wealth of general health information. The AMA's Doctor Finder indexes over 650,000 doctors of medicine (MD) and doctors of osteopathy or osteopathic medicine (DO), and provides data on each physician's medical school and residency training, year of graduation from medical school, and primary practice specialty. An office address is also provided (information on the current address, however, may not be accurate). Searchers can look for physicians by location and name, or medical specialty. The Hospital Finder provides information on the hospital's accreditation status, number of staffed beds, admissions, surgeries, and services lines such as psychiatry, cardiology, pediatrics, radiation therapy, and so on. This information is particularly important for individuals in smaller communities, where the local hospital may not provide or specialize in the specific therapy that a patient needs. The "General Health" selection bar includes an interactive "personal family health history," which the user can fill out and print as a personal medical record. The "You and Your Doctor" choice on the same selection bar includes good articles on choosing a doctor, communicating with that doctor, and getting second opinions.

The American Board of Medical Specialties <http://www.certifieddoctor.org/> CertifiedDoctor Locator Service allows the public to search by geographic area and specialty for a doctor certified by the ABMS. Requirements to be board certified include three to seven years of postgraduate medical training in an approved residency or fellowship program and successful completion of an extensive certifying test. Board certification status, address, telephone, and hospital and health plan affiliations are available online for those physicians who subscribe to the service. Basic information, including address and board certification, is available on those physicians who don't subscribe but have still been board certified in their specialties.

U.S. News and World Reports Online <http://www.usnews.com/> contains a large amount of health information for consumers, including the Best Hospitals

Finder <http://www.usnews.com/usnews/nycu/health/hosptl/tophosp.htm> and America's Top HMOs <http://www.usnews.com/usnews/issue/981005/5hmol. htm>. This site is outstanding for active health care participants who need to find top hospitals or information on choosing the best HMO plan. Links to a basic health care schedule, medical dictionary, Ask-the-Doc, and the Johns Hopkins IntelliHealth Web site are added bonuses.

ThriveOnline <http://thriveonline.com/> is not only a good health information Web site but also an excellent source for advocacy issues. "The Powerful Patient" <http://www.thriveonline.com/health/powerful/index.html> tells visitors how to get the best care with doctors, hospitals, and health plans. Legal issues concerning right to die and advance directives are covered. Five steps for the newly diagnosed patient provide advice on getting adequate information from the health care provider, researching the disease, deciding on treatment, finding support, and planning for the future.

LEARNING RIGHTS AND RESPONSIBILITIES

Just as a car buyer should know her protections under the law against unscrupulous dealerships, the consumer should know her rights and responsibilities in health care. Insurance companies and HMOs are notorious for refusing payment or treatments, and physicians may not offer treatment options or share needed information. Privacy of medical records can be critical to avoid discrimination in the work place; taking time off for family medical emergencies may jeopardize employment should the employer choose to ignore the Family and Medical Leave Act. Patients may be pushed into treatments they don't want, or their lives prolonged beyond their wishes for lack of the designation of a health care surrogate or advance directives.

The Patient's Bill of Rights <http://www.aha.org/resource/pbillofrights. html> was first adopted by the American Hospital Association in 1973, and has been expanded and updated since then to include the patient's responsibilities. Under the Joint Commission of Accreditation of Hospitals standards, a copy of the Rights must be given to every person admitted to a health care facility. It should be used by patients to understand their basic rights as well as their own responsibilities for their health. The Mental Health Patient's Bill of Rights, covering mental health and substance abuse treatment services, can be found at the American Psychological Association's Web site <http:// www.apa.org/pubinfo/rights/rights.html>.

The National Patient Advocate Foundation <http://www.npaf.org> provides political information on health legislation and policy reform, especially as it relates to patients' rights. Information on contacting national representatives, state insurance commissioners, and state attorneys general is available for those patients struggling with insurance companies. In a similar vein, the

Patient Advocate Foundation <http://www.patientadvocate.org/> provides education and legal counseling to cancer patients on managed care, insurance, and financial issues.

Consumers International, based in London, is an international group of consumer organizations with the mission of defending the rights of consumers. Its publication, "Campaigning for Patients' Rights: A Guide for Patients and Consumer Activists" <http://www.consumersinternational.org/campaigns/patientsrights/guide.html>, shows the international trend in patients' rights and is helpful for travelers in countries other than the United States.

The National Coalition of Mental Health Professionals & Consumers, Inc. <http://www.nomanagedcare.org/>, is blatantly anti-managed care. The information it provides, however, with its "Eleven Unethical Managed Care Practices Every Patient Should Know About" and "Consumer Manual," is excellent for all patients insured with an HMO.

NOAH (New York Online Access to Health) is a superb location for health information and also addresses patients' rights, advance directives, privacy issues, resources, and physician information <http://www.noah.cuny.edu/patients.html>. While the information is directed to New York residents, much is applicable to the rest of the U.S. population.

The National Coalition for Patient Rights <http://www.nationalcpr.org> directs its information to privacy issues. Information includes news articles, legislation, and strategies that patients can utilize to protect their medical records.

The Department of Labor's Web site <http://www.dol.gov/pwba/public/health.htm> has a wealth of information about federal protections, especially for working women. The Health Care Bill of Rights, Newborns' and Mothers' Health Protection Act, women's changing needs in health benefits, and other benefits information is all available here.

The Family and Medical Leave Act (FMLA) has been in effect since 1993, but is still unknown to many working women. The National Partnership For Women and Families <http://nationalpartnership.org> offers concrete information and advice for caregivers, making the most of a managed care plan, using the Act for family or medical leave, and more.

COMMUNICATION WITH HEALTH CARE PROVIDERS

Women have different communication styles and ways of reporting medical symptoms than men.[7] Many women have been culturally indoctrinated to avoid challenges with authority figures, and often will not question or challenge their physicians' statements. One study showed that women's symptoms were twice as likely to be judged to be psychosomatic as men's,[8] perhaps due to the communication and cultural differences. Since women often have difficulty being taken seriously when they voice their health concerns, they must learn effective communication techniques or bring a skilled advocate who can convey their questions, needs, and desires to the health care provider.

Due to the nature of the medium, Internet information tends to be brief and somewhat superficial (books provide a better resource for covering communications between patients and health care providers). One of the better Internet sites for communication issues is that of the Consumer Information Center in Pueblo, Colorado <http://pueblo.gsa.gov/health.htm>. Among the over 75 available health pamphlets, nine deal with questions for doctors and pharmacists and aids for keeping personal health records.

CallConnect Communications features a demonstration of the American Institute for Preventive Medicine's Health at Home <http://www.chsdemo. com/present/selfcare/main.htm>, an online book of health information. Brief articles on medical decisions, preventive health timetables, informed consent, rating health care providers, dental checkups, and more is included.

Medical Information Unlimited <http://www.medinfo4u.com> is an organization offering a number of products and services designed to help people make better health care decisions. While most of its services are fee-based, the self-care articles are available free of charge and include how to critically review medical information, choose a physician, talk to the doctor, deal with the emotional aspects of illness, and understand clinical trials.

FINDING GOOD HEALTH INFORMATION ON THE INTERNET

Searching the Internet can be a daunting prospect. Many sites provide excellent information specifically for women and can easily be found through search engines or links. Finding health information isn't as tricky as planning the search strategies, using proper terminology, and–most importantly–discerning credible information. The neophyte searcher can use "The Patient's Guide to Healthcare Information on the Internet" <http://www3.bc.sympatico. ca/me/patientsguide> as an excellent tool for learning the step-by-step process of finding dependable data. Medical terminology, databases, reference sites, search engines, and determining reliability are all well explained. Links to PubMed and reputable Web sites are included. Users should always be cautioned to use evaluation criteria, such as the Health On the Net (HON) Foundation <http://www.hon.ch> or Quackwatch <http://www.quackwatch. com>, to determine the validity of all information found on the Internet.

CONCLUSION

Patients will always need good physicians and health care systems. With the staggering statistics of medical errors, patients must be willing to accept their share of the responsibility for their own health care so that they can eliminate what errors are under their control. If the health care provider doesn't have a complete medical record, the patient should provide her own copy. If surgery is to be performed, some patients have taken to marking their own limbs to designate the correct part. Women should always obtain information on the drugs that they are receiving from their physicians, pharmacists, or hospital. The shared decision-making trends in today's medical marketplace are to all consumers' advantage in that health care providers are, for the most part, more willing to "share the power." It can be very difficult for the patient to deal effectively with the health care system, especially when she is ill, emotionally drained, and perhaps anxious about her health outcome. But without consumer skills, and the emotional strength to assert her needs and desires, the female patient faces the risk of substandard treatment and poor quality of care. The Internet provides a valuable resource for women to increase their consumer skills and health knowledge, for a better quality of life.

NOTES

1. Kohn, L.T.; Corrigan, J.M.; and Donaldson, M.S. *To Err Is Human: Building a Safer Health System*. Washington, DC: National Academy Press, 1999. Advance copy is available at <http://books.nap.edu/html/to_err_is_human/>.
2. Haslegrave, M. "What She Wants." *Lancet* 349(March 1, 1997): S1.
3. Rose, V.L. "Race, Gender Affect Which Treatment is Selected for Angina Symptoms." *American Family Physician* 60(August 1999):603.
4. Collins, K.S. et al. "Health Concerns Across a Woman's Lifespan: The Commonwealth Fund 1998 Survey of Women's Health." NY: The Commonwealth Fund, 1999. Available: <http://www.cmwf.org/programs/women/ksc_whsurvey99_332.asp>.
5. Goldstein, D., and Toth, C. "Women on the Internet: Empowered, On-line, and Making Decisions." *Managed Care Interface* 12(July 1999):40-2
6. *Networking For Better Care: Health Care In the Information Age*. Washington, DC: Benton Foundation, 1999. Available: <http://www.benton.org/Library/health/>.
7. Klemm, P. et al. "Gender Differences on Internet Cancer Support Groups." *Computers in Nursing* 17(March-April 1999):65-72.
8. Kroenke, K., and Spitzer, R.L. "Gender Differences in the Reporting of Physical and Somatoform Symptoms." *Psychosomatic Medicine* 60(1998):50-155.

Women and Physical Fitness

Diana J. Cunningham
Janet A. Ohles

ABSTRACT. Although the promotion of health, sports, and physical fitness are pervasive themes as well as part of federal U.S. policy, women lag behind their male counterparts in the areas of health and physical fitness. And, although there is a general trend toward increased participation of women in sports and physical activity across a life span, a large number of women promise like Scarlett O'Hara to "think about it tomorrow." This article highlights some demographics and results of key research, then focuses on qualitative online resources for women over age 25 that are related to women's health and physical fitness. Definition problems exist, but the terms sports and physical fitness are used interchangeably. There are several challenges in locating quality Web sources related to women, health, and sports/physical fitness. A major one is the need to "surf" from many subject areas. The print and electronic resources that are highlighted all reflect the truly diverse nature of the available materials–everything from the prosaic "how to buy running shoes" to the clinical questions relating to an association between physical activity and coronary heart disease in women. Written for women, the article is intended as a starting point for substantive informational sources largely from the Internet. *[Article copies available for a fee from The Haworth Document Delivery Service: 1-800-342-9678. E-mail address: <getinfo@haworthpressinc.com> Website: <http://www.HaworthPress.com>]*

KEYWORDS. Women, physical fitness, sports, Internet

Diana J. Cunningham, MLS (diana@nymc.edu), is Associate Dean and Director, Medical Sciences Library, New York Medical College, Valhalla, NY 10595. She is currently working on an MPH. Janet A. Ohles, MLS (Janet_Ohles@nymc.edu), is Head, Information Services, Medical Sciences Library, New York Medical College, Valhalla, NY.

[Haworth co-indexing entry note]: "Women and Physical Fitness." Cunningham, Diana J., and Janet A. Ohles. Co-published simultaneously in *Health Care on the Internet* (The Haworth Press, Inc.) Vol. 4, No. 2/3, 2000, pp. 85-98; and: *Women's Health on the Internet* (ed: M. Sandra Wood, and Janet M. Coggan) The Haworth Press, Inc., 2000, pp. 85-98. Single or multiple copies of this article are available for a fee from The Haworth Document Delivery Service [1-800-342-9678, 9:00 a.m. - 5:00 p.m. (EST). E-mail address: getinfo@haworthpressinc.com].

85

INTRODUCTION

As the year 2000 begins, American culture is characterized by empowered consumers, growth of online resources, and new efforts to promote health and fitness. The voice of the consumer is growing from a whimper to a roar.[1] The face of health care is changing from the "doctor knows best" to the consumer, who is now more likely a woman who needs to know more. A special section of a recent *Wall Street Journal* entitled "Patient, Teach Thyself" is one good example.[2] The promotion of health, sports, and fitness are pervasive themes. Though females form the majority of the American population and live longer than men, they pale in comparison to their male counterparts in the areas of health and physical fitness. Despite a general trend toward increased participation of women in sports and physical activity across a life span, a large number promise, like Scarlett O'Hara, to "think about it tomorrow." This article will highlight some demographics and results of key research, then focus on qualitative online resources for women over age twenty-five related to women's health and physical fitness or sports. Definition problems exist. Broad terms such as "physical fitness," "sports," or even "women" make searching for meaningful information difficult. The literature of sports medicine and rehabilitation is also beyond the scope of this article. This is written for women as a "jumping off" place for substantive informational sources largely from the Internet.

In general, people in the United States are living longer, and women are living the longest. The Census Bureau reports that the 1990 population by gender breakdown is 48.7% male and 51.3% female.[3] Projections for 2000 reveal this small, but significant, majority will not change for 2000 and will continue through 2050. In addition, the median age of 32.8 years continues to rise with the "graying of the population" to a median age of 35.7 by the year 2000 and 38.1 by 2050.[3] The landmark *Physical Activity and Health: A Report of the Surgeon General*[4] concludes that:

- More than 60 percent of U.S. women do not engage in the recommended amount of physical activity
- More than 25 percent of U.S. women are not active at all
- Physical inactivity is more common among women than men
- Social support from family and friends has been consistently and positively related to regular physical activity

This report highlights how physical activity impacts health, specifically the link between physical activity, morbidity, and mortality. The recommendations focus on cardio-respiratory endurance and specified sustained periods of vigorous physical activity involving large muscle groups, lasting at least twenty minutes, on three or more days per week. The report acknowledges that our highly technological society makes it increasingly convenient to remain sedentary and discourages physical activities in obvious and subtle ways. The challenge for the future is fostering sustained personal fitness, particularly for women.

With health care now a trillion-dollar business and government a major provider via Medicare and Medicaid, the U.S. has developed major initiatives to promote physical activity and health. Since 1979 the Centers for Disease Control and Prevention have promoted *Healthy People 2000,* the nation's prevention agenda.[5] "Physical activity and fitness" is one of twenty-two priority goals. As preparations near completion for *Healthy People 2010,* Surgeon General David Satcher has assessed current results: overall 15% of the objectives have been met, 44% are progressing toward their targets, but 20% are moving away from their targets.[6] Physical activity and fitness progress falls into this last objective. Multiple objectives set targets for reducing coronary heart disease deaths and overweight prevalence, for increasing the proportion of people who engage regularly in physical activity, and for engaging in vigorous physical activity.[7,8] Though included by implication, special targets for women are named in relation to race, ethnicity, low-income, and some age groups. Adult women, as an aggregated grouping (aged 25 and over) are not singled out. Full text of both initiatives, including midcourse reviews and results, is available via <http://web.health.gov/healthypeople/>. The prestigious Institute of Medicine has also released its final report entitled *Leading Health Indicators for Healthy People 2010,* available at <http://odphp.osophs. dhhs.gov/pubs>. Promotion of physical activity is a national priority.

The story of women and sports is a fascinating, but complex, tale that is beyond the scope of this article. However, results from a few selected studies will be presented. Patricia Vertinsky in her published speech with the telling title "Run, Jane Run . . ." provides an extraordinary review of the "central tensions" behind women who exercise and summarizes why so few women are active enough to benefit their own health, despite the increasing evidence.[9] Notwithstanding repeated and extensive government campaigns to educate the public, she reports that fewer women than men participate in every age group. Vertinsky concludes, "Something is drastically wrong when exercise is said to be associated with so many health benefits, yet only a small portion of the female population exercises sufficiently to accrue these benefits."[9] Eyler and her gifted team provide an excellent overview in "Physical Activity and Women in the United States: An Overview of Health Benefits, Prevalence, and Intervention Opportunities."[10] This study confirms that even with decades of physical activity research and interventions conducted on men, very little is known about the patterns of physical activity among U.S. women. Many of the published studies have reportedly been conducted solely on men.[10] For instance, cardiovascular disease and coronary heart disease have been studied extensively in men, but not in women. Further, the Surgeon General's report on *Physical Activity and Health* finds that only 8 of 55 population-based studies included women.[4] What is known, however, is that women who are active have more favorable cardiovascular risk profiles than inactive women, improved blood pressure, improved body mass index, and better blood lipid concentrations.

Vigorous physical activity by women is generally not prevalent–even today. There are some historic reasons. As recently as the nineteenth century, middle-class women who exercised were deemed unladylike. "Well-to-do women may have played golf or tennis, but they played while corseted and gloved."[11] Lower-class women had heavy domestic chores that were physically demanding and stressful. Even the current generations of women were discouraged from being physically active so as not to damage their reproductive organs or jeopardize future fertility.

Old myths and prejudices about women and physical fitness take a long time to change. Lutter writes one of the very best histories of women in sports in an issue of *Clinics in Sports Medicine* that was devoted to the athletic woman.[12] Sallis reports that if women do not have a history of physical activity at an early age, they are less likely to be physically active as adults.[13] Levels of sporting activity are predicted by previous sports participation at school and in early life. Of those who are active, Sternfeld et al. studied physical activity patterns by specific domains and concluded that "women with the highest level of participation in sports/exercise and active-living behaviors were more likely to be younger, white, college-educated, without young children at home, and leaner."[14] Women with the highest level of household/care-giving activity were more likely to be older, Hispanic, married with young children at home, unemployed, and perceive that they have little time for exercise. Masse recently reported findings of 53 experts brought together under the Women's Health Initiative of the National Institutes of Health to identify important issues related to measuring physical activity in minority women, women in mid-life and older (40+ years of age).[15] One of their concerns was that current survey instruments are not specific to activities performed by women, especially those who are not of European descent.[15] Finally, though Title IX, the Education Amendments Act of 1972, was passed to encourage and safeguard equal opportunity in sports for women and girls in college programs that received federal money, 90% of schools are not yet in compliance with the 1972 law.[16]

Part of the success of the 1971 publication *Our Bodies, Ourselves* was that it served as a benchmark for women taking responsibility for their own health and fitness.[17] What began as a small women's discussion group in Boston calling themselves "the doctors group" and reacting against feelings of frustration and anger towards specific doctors and the medical maze, has grown into the Boston Women's Health Book Collective. It continues to promote women's health and fitness, e.g., *Our Bodies, Ourselves for the New Century,* with a special chapter on online women's health resources.[11] Further, Title IX of the 1972 Education Act greatly increased athletic opportunities for college women, but the mothers of these women may have already negatively influenced their daughters. In sum, U.S. women age 25 and over form a large and special cohort that must embrace regular and vigorous physical activity for their own health. Providing solid Web searching pointers and qualitative Web sites may provide a new beginning.

GENERAL WEB SITES

Many health care and medical associations, federal offices, medical journals, and popular publications provide quality information on the World Wide Web about women, in relation to physical fitness and health. For purposes of this article, the terms "physical fitness" and "sports" are used interchangeably, recognizing the oversimplification. The following Web sites are good places to start.

American Orthopaedic Society for Sports Medicine (AOSSM) <http://www.sportsmed.org/>

The AOSSM was formed in 1972 with its primary purpose to serve as a forum for education and research. Membership is composed of orthopedic surgeons, specializing in sports medicine. Members must spend a minimum of sixty percent of their practice on sports medicine and be involved in medical coverage of a sports team, either professional, academic, or community-related.

The "Publications and Products" section of the Web site has a patient education link. "Sports Tips" provides instruction sheets on a variety of topics from exercising for bone health to designing a weight training program. "Sports Tips" may be read online, downloaded as a Word document, or ordered for a small fee in bulk. Documents on preventing sports injuries and the importance of physical activity for mentally retarded individuals are also online. In addition, the "Professional" section, under the publications and products link, has very specific information, e.g., how to remove a football helmet in the event of an injury, a monograph on concussion in sports that may be downloaded, or summaries of monographs that may be ordered. The site is very easy to navigate with information of a high quality.

Centers for Disease Control and Prevention (CDC)
<http://www.cdc.gov/health/physact.htm>

The CDC is an agency of the Department of Health and Human Services, with a mission to promote health and quality of life by preventing and controlling disease, injury, and disability. Within the CDC there are eleven centers, offices, and institutes of which one is the National Center for Chronic Disease Prevention and Health Promotion (NCCDPHP). The NCCDPHP programs and activities address the impact of arthritis, cancer, cardiovascular disease, diabetes, and oral diseases and conditions, along with the relationship of tobacco use, physical activity, and nutrition to chronic disease conditions.

In addition, the NCCDPH has an abundance of information from the "Health Topics A-Z/Physical Fitness and Health" link on their Web site. There are three main areas covered: "Nutrition and Physical Activity," "Physical Activity," and full text of the Surgeon General's Report (noted above). The variety of documents is impressive. For example, under the "Nutrition and Physical Activity" Nutrition link is the "Physical activity. Ready. Set. It's everywhere you go!" kit. The resources and materials in this kit are designed to assist the busy adult in working physical activity into their schedules. Anyone interested in the topic of physical fitness may first want to review the materials at the CDC's Web site. This one is for everyone!

Feminist Majority Foundation Online
<http://www.feminist.org>

The Feminist Majority Foundation was established during 1987 to 1989. The Foundation defines its mission to create research, educational programs, and strategies to further women's equality and empowerment, to reduce violence toward women, to increase the health and economic well-being of women, and to eliminate discrimination of all kinds. This well-organized Web site with descriptive buttons allows one to locate information with ease. The "Women & Girls in Sports" button leads one to interesting documents such as "Empowering Women in Sports" and "Sports and the Disabled." Each document contains references, statistical data, and is well written. The links to other "Internet Resources" have annotated narratives and are the *most comprehensive* found.

JAMA Women's Health Information Center
<http://www.ama-assn.org/special/womh/womh.htm>

The *JAMA* Women's Health Information Center is produced and maintained by the American Medical Association, with a stated audience of physicians and other health professionals. *JAMA* editors and staff are responsible for the content, and a biographical sketch is provided for each member on the editorial review board. The site indexes several sources with articles from *JAMA* and abstracts from other journals pertinent to terms entered. Abstracts are also provided from major newspapers, such as the *Boston Globe,* with hypertext links to the full text of the article at the newspaper's Web site.

The home page has a search button that does *not* require Boolean logic, but uses the Excite search engine that retrieves articles by matching keyword prefixes and a concept approach. Entering the word "sports" in the search box retrieves many articles dealing with sports issues. It also retrieves many false hits from keyword matches. Entering the phrase "sports physical activity" retrieves a set of higher accuracy. One drawback to the site is that the only topics highlighted from the home page are "sexually transmitted diseases" and "contraception." Additional topics are covered under a newsline service from Reuters, giving recent top stories, and special reports from major professional sources that are not defined. Overall, the site retrieves excellent information when conducting a search, but could benefit from presenting users with a search option by topics and offering a list of sources.

National Women's Health Information Center (NWHIC)
<http://www.4woman.org/>

NWHIC is a service of the Office on Women's Health in the U.S. Department of Health and Human Services. The Web site is designed to provide information to advance women's health research, services, and public and health professional education. NWHIC takes a lead for the women's section of *healthfinder* <http://www.healthfinder.gov>, the free gateway to reliable consumer health information, also developed by the U.S. Department of Health and Human Services.

The NWHIC Web site is easy to navigate, with a "Search By Health Topic" button on the home page. Clicking on the letter "s" from the "Topics" page links to a dialog box that has three sports-related topics: "Sports and Recreation," "Sports Injury," and "Sports Medicine." Information is presented on publications or organizations in each sports-related topic. Also available are online tip sheets written by professional associations (e.g., American Academy of Orthopedic Surgeons) and reprints of articles originally published in established consumer health publications, such as the *FDA Consumer.* Additional links lead one to listings of national and international organizations (e.g., Women's Sports Foundation and Special Olympics International), along with information on governmental councils, like the President's Council on Physical Fitness and Sports.

Shape Up America! Fitness Center
<http://www.shapeup.org/>

Shape Up America!, founded by former Surgeon General Dr. C. Everett Koop, is a national initiative to promote both healthy weight and increase physical activity. The three primary objectives are: (1) to promote a new understanding by Americans of the importance of achieving and maintaining a healthy weight and increasing physical activity, (2) to inform Americans of the logical, proven ways to achieve a healthy body weight, and (3) to increase cooperation among national and community organizations committed to advancing healthy weight and increased physical activity as major public health priorities. The Web site provides a superb section on nutrition and physical fitness, and presents weight and activity information not found on other Internet sites.

A "Body Mass Index" (BMI) page not only defines BMI, describes its importance and how to measure one's BMI, but also calculates body mass index online. The "Fitness Center" pages have five assessment tests: Par-Q, Activity, Flexibility, Strength, and Aerobic. A "Nutrition" page explains the daily caloric goal, the food pyramid chart, and includes menu plans for 1800, 2000, and 2500 calories. "Improvement" pages have four focus areas designed to meet the needs of individuals with sedentary jobs, those who are trying to become more active, those who want to commit to a vigorous exercise program, or those who want to increase their muscle amount. A "Barriers" page lists common obstacles to beginning an exercise program (e.g., it's too cold outside), and a solution to the barrier. Shape Up America! is a wonderful location for anyone who wants to develop an exercise program, and also is a reference source for those already involved in sports.

The *New York Times* Women's Health
<http://www.nyt.com/women/>

The *New York Times* Web site includes a first-rate section on Women's Health, with "Diet and exercise" offered as a topic selection. Each topic area has selected articles from The *New York Times*, along with links to other online resources. A filter to the BarnesandNoble.com Web site offers an efficient means to locate consumer health and personal interest titles. The Web site has an extensive listing of online resources organized by topical areas, making it easy for one to keep up-to-date on new Web sites. This site is free of charge, but users must register.

GENERAL PHYSICAL FITNESS/SPORTS WEB SITES

American Running Association
<http://www.americanrunning.org/>

The American Running Association was established in 1968 and is dedicated to providing programs and information on fitness, training, nutrition, and injury prevention, treatment, and rehabilitation. The Association publishes *Conquering Athletic Injuries* and the monthly *Running & FitNews.* Members include individual runners, exercise enthusiasts, and sports medicine professionals.

The Web site offers a special feature, "Women's Health & Fitness," with both condensed articles from peer-reviewed medical journals, and articles written by reputable medical staff. While there were only eight articles provided in the "Women's Health & Fitness" section in December 1999, additional topics are located in other sections, such as "Sports Psychology," "Fluids & Nutrition," and "Weight Management." The Web site complements the diversity found in the membership and has valuable information *beyond* that relevant to individuals involved in running.

Dr. Pribut's Sports Page
<http://www.clark.net/pub/pribut/spsport.html>

Dr. Pribut, Clinical Assistant Professor of Surgery in the Department of Surgery at George Washington University Medical Center, is Board Certified in Foot Surgery by the American Board of Podiatric Surgery and in Foot Orthopedics by the American Board of Podiatric Orthopedics and Primary Podiatric Medicine. His Sports Page is dedicated to providing useful information to patients on foot health matters.

Although the focus is on the foot, the adage is "when your feet hurt you hurt all over." While Dr. Pribut provides comprehensive information on foot health (e.g., choosing a running shoe and other health matters), the pages go beyond foot health (e.g., "Selecting a Sports Physician" or "Women's Health and Sports"). Also available are " Dr. Pribut's Internet Columns from the American Podiatric Medical Association News," and summertime Web specials–among other documents. This Web site is a must!

HealthLink
<http://healthlink.mcw.edu/sports-medicine/>

A multidisciplinary team of physicians and medical informatics specialists creates "HealthLink from the Medical College of Wisconsin Physicians &

Clinics." HealthLink is the strongest academic site found on women and physical fitness, although by their own definition, the general public is not their primary clientele. HealthLink lists three goals for their Web site: (1) to promote the health and well being of HealthLink's community; (2) to empower HealthLink's patients with accurate, timely, impartial, and authoritative medical information; and (3) to strengthen the sense of community among HealthLink's patients and physicians.

The sports medicine Web page is not limited to the female fitness activist, but provides information pertinent to all ages, gender, and races. Each article is written in a consumer health style, with links to related articles. Articles cover a breadth of topics, such as teen growing pains, gait and motion analysis, foot and ankle surgery, and tendinitis. HealthLink editors construct quality pages, giving background on the authors, presenting contact information, and citing the peer reviewed literature, when appropriate. This is a well-documented source.

Physical Activity & Health Network (PAHNet)
<http://www.pitt.edu/~pahnet/>

PAHNet is a collaborative effort developed by individuals at the University of Pittsburgh, the Children's Hospital of Boston, and the Children's National Medical Center. A key feature of the site is translations in German, French, Italian, Spanish, and Portuguese. PAHNet states their mission is to consolidate information on the health implications of physical activity and exercise by connecting individuals that work in this discipline and facilitate communication among them.

The Web site offers comprehensive information on physical fitness presented in an easy-to-use manner with sections on "Recommendations" (e.g., "Physical Activity and Cardiovascular Health," a consensus development conference statement from the National Institutes of Health); "Promotion Programs" (e.g., "Guidelines for School and Community Programs to Promote Physical Activity" which is published in the *Morbidity and Mortality Weekly Report*); and "Epidemiology" (e.g., *NIH Bibliography on Physical Activity and Cardiovascular Health*). Other subject areas are "Health-Related Physical Fitness Sites," giving online links to Web sites; "Organizations," giving a comprehensive list of international, national, and professional organizations that have sports-related information; and "Listservs." PAHNet is a solid and comprehensive source.

The Physician and SportsMedicine/Online
<http://www.physsportsmed.com/>

The Physician and Sportsmedicine is a peer-reviewed clinical sports medicine journal published monthly, serving the practicing physician's professional and personal interests in the medical aspects of exercise, sports, and fitness. The journal covers practical, primary care-oriented topics such as diagnosing and treating knee and ankle injuries, managing chronic disease, preventing and managing overuse injuries, helping patients lose weight safely, and exercise and nutrition topics.

The Web site is extremely well organized and comprehensive in coverage. The "Personal Health" link offers access to a variety of well-organized topics, such as women's health, sports equipment and apparel, and rehabilitation. The "Resource Center" link provides all-inclusive information on directories, associations, a pre-participation physical evaluation, and links to other Web sites.

WOMEN'S PHYSICAL FITNESS/SPORTS WEB SITES

Canadian Association for the Advancement of Women and Sport and Physical Activity (CAAWS)
<http://www.caaws.ca/>

Founded as a nonprofit organization in 1981, the mission of CAAWS is to ensure that girls and women have access to a complete range of opportunities, choices, and equity as participants and leaders in sport and physical activity. CAAWS facilitates many partnerships to advance women and sports with other organizations, including the Canadian Centre for Ethics and Sports, With Sport Canada, Canadian Olympic Association, and Coaching Association of Canada.

The Web site has a health section with articles covering a wide variety of information including concussion injuries, medical problems associated with the female athlete triad, exercise and its ability to help one stop smoking, and knee injuries. Additional Web sections provide books, brochures, and pamphlets available for purchase; a section on women and sports history; gender equity; and harassment. The Web site has an abundance of exceptional online information.

Melpomene
<http://www.melpomene.org/>

Melpomene was founded in 1982 with a mission to help girls and women of all ages link physical activity and health through research, publication, and education. Membership includes those professionally trained in health care, physical activity, and sports for girls and women. Melpomene conducts research and disseminates information on issues such as body image, osteoporosis, athletic amenorrhea, exercise and pregnancy, and aging. The organization publishes *The Melpomene Journal,* which includes reports on current research, along with general interest articles and profiles.

The Web site presents listings of women's sports associations, a library that includes "Tip sheets" (e.g., "Exercise and Women of Color"), and a "What's New" section that describes the latest research. A summary of the Cigna report on physical activity for women in a work setting is included in the "Latest Research" section, along with information on *The Melpomene Journal.* "What's New" also lists articles where Melpomene is cited, highlighting women and sports issues. In December of 1999, the "What's New" section included references, such as Lutter, J.M. et al., in a recent article entitled, "Incentives and Barriers to Physical Activity for Working Women."[18] Unfortunately, full bibliographic citations are *not* offered, and many of the journals are not well indexed in health science databases. The Web site also includes an online catalog where one can purchase books, such as *The Bodywise Woman,* and brochures, such as "Choosing Good Books About Girls in Sports," "Girls, Self-Esteem & Sports," and "Selecting a Prenatal Exercise Program." Overall, Melpomene has high quality information with resource listings that include superior online links to other Web sites, books, journal articles, and other organizations.

The Women's Sports Foundation
<http://www.lifetimetv.com/WoSport/index.htm>

Tennis's Billie Jean King and other champion female athletes established the Women's Sports Foundation in 1974. Membership is open to anyone interested in women's sports. The Foundation's goal is to improve the physical, mental, and emotional well-being of women through sports and fitness participation. There are four program areas for the consumer: education, opportunity, advocacy, and recognition. A "Health & fitness" area has excellent content, although the pages are very busy and cluttered in appearance, making it difficult to find information. Within "Health & fitness," there is a dialog box where one can search for specific topics including sports. The "Sports" pages have a variety of information on how to find a personal trainer, directions on exercises to strengthen and tone specific areas of the body, written by workout experts, and instructions on how to exercise without injuries, among other topics. Though clumsy to search, Women's Sports Foundation provides useful information.

FINAL THOUGHTS

Many Web sites devoted to women's issues contain information on sports and physical fitness. Some documents also pertain to health care. As a dynamic resource, there can be no static or comprehensive list of Web sites. Health care in relation to women's physical fitness or sports may be located on a variety of sites devoted to women, sports, news coverage, or health. It is important to mention that the sites chosen for this article are not listed on any *one* Web site, index or monograph, but are a selected compilation from many sources. Further, qualitative Web criteria from <http://library.nymc.edu> was used as a framework to assess the sites. To obtain current information on health care issues for women and sports, many Web sites and indexes should be searched. And, print resources, such as *Our Bodies, Ourselves for the New Century,* cannot be overlooked, since they can, paradoxically, provide good World Wide Web information. Also, little overlap was apparent in researching sources to newer Web resources.

Health and physical fitness are clear national priorities. There is every reason to promote life style changes and fund research that better understands interventions needed to encourage more adult women to exercise. The April 1994 volume of *Clinics in Sports Medicine* is devoted to the athletic woman. Rosemary Agostini, M.D., guest editor for this special research issue, notes that it "will ask more questions than give answers."[19] The Web sites highlighted in this article are selected to provide a beginning point. *All* demonstrate the popularity of sports and physical fitness and are designed to cover the range from the very general (Shape Up America! Fitness Center) to the very specific (the Women's Sports Foundation). Searching is a process, and no one Web site will be enough. Oversimplification is dangerous. Any one of the sites noted, or references cited, can serve as step one. Scarlett, today is the day!

NOTES

1. Golodner, L. F. "Consumer Voice." In: Cohen, E.G., and DeBack, V. *The Outcomes Mandate: Case Management in Health Care Today.* St. Louis: Mosby, 1999.

2. Rout, L., ed.. "Patient, Heal Thyself." *Wall Street Journal Reports* 232 (October 19, 1998, Section R):R1-28.

3. U.S. Bureau of the Census. *Statistical Abstract of the U.S.: 1998.* Washington, DC: GPO, 1998.

4. U.S. Health and Human Services. *Physical Activity and Health: A Report of the Surgeon General.* Atlanta, GA: U.S. Department of Health and Human Services, Centers for Disease Control and Prevention, 1996. Also available online at <http://www.cdc.gov/nccdphp/sgr/npai.htm>.

5. U.S. Health and Human Services. *Healthy People 2000.* Washington, DC: U.S. Department of Health and Human Services, 1990. Also available: <http://www.health.gov/>.

6. Satcher, D. "A Decade of Many Gains in Americans' Health." *Academic medicine* 74 (July 1999):808-9.

7. U.S. Health and Human Services. *Healthy People 2010*. Washington, DC: U.S. Department of Health and human services, 1999. [Online]. Available: <http://www.health.gov/>.

8. U.S. Health and Human Services. *Healthy People 2000 Review, 1998-1999*. Hyattsville, MD: DHHS, 1999.

9. Vertinsky, P. "'Run, Jane, Run': Central Tensions in the Current Debate about Enhancing Women's Health Through Exercise." *Women & Health* 27(4, 1998): 81-111.

10. Eyler, A.A.; Brownson, R.C.; King, A.C. et al. "Physical Activity and Women in the United States: An Overview of Health Benefits, Prevalence, and Intervention Opportunities." *Women & Health* 26(3, 1997):27-49.

11. Boston Women's Health Book Collective. *Our Bodies, Ourselves for the New Century*. New York: Touchstone, 1998.

12. Lutter, J.M. "History of Women in Sports: Societal Issues." *Clinics in Sports Medicine* 13(April 1994):263-79.

13. Sallis, J.F.; Hovell, M.F.; and Hofstetter, C.R. "Predictors of Adoption and Maintenance of Vigorous Physical Activity in Men and Women." *Preventive Medicine* 21(1992):237-57.

14. Sternfeld, B.; Ainsworth, B.E.; and Quesenberry, C.P. "Physical Activity Patterns in a Diverse Population of Women." *Preventive Medicine* 28 (March 1999):313-23.

15. Masse, L.C.; Ainsworth, B.E.; Tortolero, S. et al. "Measuring Physical Activity in Midlife, Older, and Minority Women: Issues from an Expert Panel." *Journal of Women's Health* 7 (February 1998):57-67.

16. Mosley, B.F. "Entitled by Title IX: The Law that Secures a Woman's Right to Play Celebrates its Silver Anniversary." *Women's Sports and Fitness* 19(June 1997):28.

17. Boston Women's Health Book Collective. *Our Bodies, Ourselves: A Book by and for Women*. New York: Simon and Schuster, 1971.

18. Lutter, J.M.; Jaffee, L.; Rex, J. et al. "Incentives and Barriers to Physical Activity for Working Women." *American Journal of Health Promotion* 13(4, 1999):215-8.

19. Agnostini, R. "Preface." *Clinics in Sports Medicine* 13(April 1994): xi-xii.

The Whole Nine Months and Then Some: Pregnancy, Childbirth, and Early Parenting Resources on the Internet

Janet A. Crum

ABSTRACT. Having a baby is one of life's greatest changes, causing even the most knowledgeable health care consumers to ask many questions about pregnancy, birth, and infant health. The Internet offers many resources for prospective and new parents from many different types of sites. This article describes the most useful sites in the following categories: medical megasites–huge repositories of health information for consumers; pregnancy and parenting sites; and pediatric sites; plus specialized sites on infertility, locating a practitioner, prenatal testing, pregnancy complications, breastfeeding, postpartum depression, pregnancy loss, and more. *[Article copies available for a fee from The Haworth Document Delivery Service: 1-800-342-9678. E-mail address: <getinfo@haworthpressinc.com> Website: <http://www.HaworthPress.com>]*

KEYWORDS. Pregnancy, childbirth, parenting, Internet

The birth of a child is one of the most significant life changes a person can undergo. In addition, for many women, pregnancy represents their first sig-

Janet A. Crum received her MLS from the University of Washington in 1992. She is currently Bibliographic and Database Services Librarian, Oregon Health Sciences University, P.O. Box 573, Portland, OR 97207-0573, and the mother of an 18-month-old son.

[Haworth co-indexing entry note]: "The Whole Nine Months and Then Some: Pregnancy, Childbirth, and Early Parenting Resources on the Internet." Crum, Janet A. Co-published simultaneously in *Health Care on the Internet* (The Haworth Press, Inc.) Vol. 4, No. 2/3, 2000, pp. 99-111; and: *Women's Health on the Internet* (ed: M. Sandra Wood, and Janet M. Coggan) The Haworth Press, Inc., 2000, pp. 99-111. Single or multiple copies of this article are available for a fee from The Haworth Document Delivery Service [1-800-342-9678, 9:00 a.m. - 5:00 p.m. (EST). E-mail address: getinfo@haworthpressinc.com].

99

nificant contact with doctors and hospitals. There are physical exams, prenatal tests, and lots of questions, especially for women whose pregnancies are classified as complicated or high-risk. As they cope with the exciting but confusing and sometimes frightening experiences of pregnancy, childbirth, and early parenthood, many women will turn to the Internet for information and support. This article is intended to present the best sites for these women, in four general categories–medical megasites, pregnancy and parenting sites, specialized sites, and pediatric sites.

MEDICAL MEGASITES

A medical megasite, as defined for this article, is a large Web site with information on a variety of medical topics. There are quite a few of these on the Internet, both commercial and noncommercial. Those listed here are but a small percentage, selected because they cover pregnancy, childbirth, and early parenting especially well or because they offer unique features or resources.

Commercial Sites

Large commercial sites typically function as medical encyclopedias, covering a wide range of medical topics in general terms, without a lot of depth. They are usually funded through banner advertising and/or partnering with other commercial Web sites. They typically include mainstream medical information, though some have sections devoted to alternative health care. Most offer some customized or interactive features, e.g., weekly e-mail newsletters, personalized login pages, chats, and message boards, for users who register. Registration is free on all sites included in this article.

The best of the lot is InteliHealth <http://www.intelihealth.com>, a joint venture of Aetna U.S. Healthcare and Johns Hopkins University. In addition to articles on a variety of topics related to pregnancy, InteliHealth includes ultrasound videos of fetal behavior and articles on breastfeeding and infant growth and development. For materials written at lower literacy levels, try Health-Center.Com <http://new.health-center.com/>, a physician-owned company offering fairly comprehensive general information on maternal and fetal development, wellness, prenatal exams, and newborn care, feeding and general health concerns. Among the best of the rest is DrKoop.Com <http://www.drkoop.com/>, a well-known medical site headed by former U.S. Surgeon General C. Everett Koop. It offers a medical encyclopedia, feature articles, and good basic information on pregnancy, birth, and children's health. Also worth a look is CBS HealthWatch <http://healthwatch.

medscape.com/>, a consumer health site operated by Medscape. The best information on this site is available only to registered users, such as an "ask the expert" feature and articles labeled "What My Doctor Reads" that highlight relevant material from the medical literature.

Noncommercial Sites

Unlike the commercial megasites, noncommercial sites typically do not offer customizable or interactive features. Instead, they usually offer no-nonsense articles and links, representing the traditional medical point of view, though a few include alternative health information. The best of these are New York Online Access to Health (NOAH) <http://www.noah.cuny.edu/> and MEDLINE*plus* <http://nlm.nih.gov/medlineplus/>. A collaborative project of the City University of New York, the Metropolitan New York Library Council, the New York Academy of Medicine, and the New York Public Library, NOAH offers a page of pregnancy-related links in a clear, well-arranged list by subtopic. Links to materials on postpartum and infant care are included, but NOAH's greatest strength is pregnancy information. MEDLINE*plus,* from the National Library of Medicine, allows users to browse by category or search by keyword. It includes links to information sheets, some in Spanish, from government agencies and professional associations, plus pre-designed MEDLINE searches that produce recent search results from PubMed <http://ncbi.nlm.nih.gov/PubMed/> on a variety of topics. It is an excellent source for recent research and basic information on pregnancy, labor and delivery, child health, and genetic disorders.

In addition to MEDLINE*plus,* two good sources of information from the U.S. government and professional associations are healthfinder <http://www.healthfinder.gov/> and the National Women's Health Information Center (NWHIC) <http://www.4woman.org/>. healthfinder, from the U.S. Department of Health and Human Services, includes sections for pregnancy, infants, and children, plus a great search engine that easily locates materials on specific topics or conditions. The NWHIC, from the U.S. Department of Health and Human Services Office of Women's Health, features low-literacy and Spanish materials, contact information for related organizations, and links for women of color and women with disabilities, as well as a good supply of links to government and association information.

PREGNANCY AND PARENTING SITES

The Internet contains a great variety of sites related to pregnancy and parenting, from huge commercial sites to personal home pages, from traditional medical information to naturopathy, home birth, and other alternative views.

Commercial Sites

Set up at least in part to take advantage of the spending spree that usually accompanies pregnancy and parenthood, the commercial sites generally support themselves by selling maternity and baby goods in online stores and/or through partnerships with other Web retailers. The sites vary in quality, with some including a large amount of medical and baby care information and others consisting mostly of baby names and other non-medical content. Most offer additional features, such as message boards, chat rooms, and e-mail newsletters, for users who register. This section includes those sites with the best health-related content; some other features are noted, where applicable.

A number of commercial sites focus primarily on pregnancy and babies. The best of the pregnancy and baby sites is Baby Center <http://www.babycenter. com/>. It offers a menu of information in four sections–Preconception, Pregnancy, Baby, and Toddler–plus communities for people in special situations (e.g., pregnant women on bed rest, parents of preemies). Registered users can get a customized startup page and e-mail newsletters based on the baby's due/birth date. The site offers good basic information with a personal touch and a sense of community. Also, unlike many commercial sites, Baby Center acknowledges controversy and alternative points of view, where appropriate. About.Com: Pregnancy and Birth <http://pregnancy.about.com/health/pregnancy> is an online community coordinated by a childbirth educator and doula (professional labor support person). The site includes newsletters, a week-by-week guide to maternal and fetal development in pregnancy, and a list of topics with links to articles on the site and elsewhere on the Web. Online chats and lists of users by due date are available to registered users. BabyData.Com <http://www.babydata.com/> includes lots of information and links on many topics from preconception to postpartum, plus a personalized pregnancy calendar and online pregnancy diary for registered users. Of particular note is a section on alternative medicine and pregnancy and a great set of links on the pros and cons of circumcision. An extensive collection of information on pregnancy and childbirth is available at StorkNet <http://www.storknet. org/>. Features include a week-by-week pregnancy guide, message boards and chat rooms, birth stories, and lots of articles. BabySoon.Com <http://www.babysoon.com/> offers the usual array of articles, online forums, and weekly e-mails, plus two outstanding features: a week-by-week pregnancy guide with information on fetal development and a to-do list for mom, and an excellent slide show illustrating fetal development.

Another group of commercial sites offer material on parenting children of all ages, in addition to information on pregnancy and infant care. The best of these is ParentsPlace.Com <http://www.parentsplace.com/>, a division of iVillage <http://www.ivillage.com/>. It features departments for fertility, pregnancy, and children by age group, chats, and message boards. The site includes good articles on many medical conditions and a terrific "Ages and Stages" area that allows a user to retrieve relevant articles based on a specified age and area of development (e.g., physical, cognitive). Also worth a look is ParentTime <http://www.parenttime.com/>, part of Time Warner's giant Pathfinder site <http://www.pathfinder.com/>. It offers menus for pregnancy and children of various age groups from infant to teenager, plus articles from *Parenting* and *Baby Talk* magazines and a panel of experts available to answer questions submitted via email. Parent Soup <http://www.parentsoup.com/> offers similar choices, but with less medical information. Its most noteworthy feature is a large panel of experts, several of whom are available for online chats.

Finally, one smaller commercial site is worth noting. The Merck Manual Home Edition: Women's Health Issues <http://www.merck.com/pubs/mmanual_home/sec22.htm> contains the complete text of section 22 (Women's Health Issues) of the printed *Merck Manual–Home Edition*, a respected medical publication for consumers. Included are chapters on the female reproductive system, hormones, infertility, genetic testing, pregnancy and its complications, labor and delivery and their complications, the postpartum period, and more. The text is clearly written, with medical terms defined, and a variety of options are presented. This site provides an excellent, thorough overview for the layperson.

Noncommercial Sites

The commercial sites discussed in the previous section usually present the standard medical point of view, rarely addressing alternative birth practices or treatments. For these viewpoints, plus information on natural childbirth, midwifery, the family bed, and similar issues, one may turn to some noncommercial and/or personal sites, of which there are many. This section lists a few of the best, plus some starting points for finding others. Some of the following sites are not produced or reviewed by medical experts. While one should always approach medical information on the Internet cautiously, careful judgment is especially necessary for sites that lack medical oversight or other forms of quality control. That said, some of these sites do contain valuable information or may lead the reader to sites that do, while others may offer support and comfort.

Childbirth.Org <http://www.childbirth.org/> is a superb site offering a wealth of information on pregnancy, childbirth, and the postpartum period. Written by nurses, childbirth educators, doulas, midwives, and lactation consultants, the articles are somewhat slanted toward natural options and away from medical intervention. However, the site includes much information on medical interventions, plus a great collection of birth stories chronicling a wide variety of birth experiences, such as home births, hospital births with and without pain relief, Caesarian sections, complicated births, etc. Women looking for an alternative to hospital birth should also check out Birth Centers Online <http://www.birthcenters.org/>, a site created by the National Association of Childbearing Centers. This site's most valuable features are its searchable directory of birth centers and detailed introduction to the birth center concept. Also available are FAQ's on childbearing and articles on various aspects of the birth experience, most written by doctors or nurses.

The Visible Embryo <http://www.visembryo.com/>, sponsored by the National Institutes of Child Health and Human Development, is a stunning online atlas of fetal development, with detailed descriptions to accompany the wealth of images. The site illustrates twenty-three stages of development in the first trimester, then every two weeks of the second and third trimesters. It is perfect for any parent-to-be who wonders what is going on "in there."

For information on nearly every aspect of pregnancy and early parenting, check the misc.kids FAQ's <http://www.cs.uu.nl/wais/html/na-dir/misc-kids/.html>, which present the collective wisdom of posters to the misc.kids family of Usenet newsgroups. This archive includes documents on allergy and asthma, baby proofing, pregnancy and childcare books, breastfeeding, children's software, crib and cradle safety regulations, miscarriage, pregnancy (with subtopics on prenatal screening, nutrition, anesthesia, etc.), sudden infant death syndrome, vaccinations, traveling with children, and more. The documents include a range of opinions and experiences not readily available elsewhere and serve as a good introduction to the misc.kids newsgroups, which may also be helpful. However, the information comes from a variety of sources and is often personal opinion, so accuracy cannot be guaranteed.

Having a Baby on the Net: 2000 Pregnancy and Childbirth Sites <http://whatsonthe.net/pregnancymks.htm> is exactly what the name implies–a list of links to medical, commercial, noncommercial, and personal Web sites or documents related to pregnancy and childbirth. There are no annotations and apparently no serious attempt at quality control, but the list includes some good links and would be a good place for users to go when they cannot find what they need elsewhere. The same can be said of the WWW Pregnancy Ring <http://www.fensende.com/Users/swnymph/Ring.html>, a Web ring of 148 sites related to pregnancy and childbirth. Many of the sites emphasize natural childbirth, alternative medicine, and/or attachment parenting. A Web ring for home births is also available–the Home Birth Ring <http://www.geocities.com/Heartland/Hills/2510/joinring.html>, which contains 54 sites related to giving birth at home.

Many women seek support during pregnancy and the postpartum period, sometimes due to pregnancy complications or a new baby's health problems. In addition to the message boards and chat rooms found in many sites already described, one may also wish to visit Email Lists About Women's Health <http://research.umbc.edu/~korenman/wmst/f_hlth.html>, a list of over 100 mailing lists on a variety of women's health topics, including breastfeeding, Caesarian sections, and infertility. One can also find support from one of the many support groups that meet around the country. To locate one, visit the Self-Help Sourcebook Online <http://mentalhelp.net/selfhelp/>, a searchable database of over 800 self-help groups around the world. It includes groups related to various genetic disorders and birth defects, postpartum depression, twins and multiples, and much more.

SPECIALIZED SITES

The large medical and pregnancy/parenting sites described thus far contain a wealth of information on many topics. Sometimes, however, one needs more detailed information than is found in the megasites. In many cases, such information can be found in one of the many specialized sites devoted to various aspects of the pregnancy and parenthood adventure. A selection of these sites is presented here.

Women having trouble conceiving a child can find helpful information from the American Society for Reproductive Medicine <http://www.asrm.org/>. The site includes a searchable directory of members, plus FAQ's on infertility, information on selecting an in-vitro fertilization program, and fact sheets on a variety of infertility-related topics, such as genetic screening, counseling, and in-vitro fertilization. FertilityPlus <http://www.pinelandpress.com/toc.html> consists primarily of FAQ's for various fertility related Usenet newsgroups. The quality of the information varies, but the quantity is great. Additional sites related to infertility can be found via the Infertility Ring <http://www.geocities.com/Heartland/Ranch/3899/ring.html>, a Web ring of thirty-seven personal pages, online support groups, commercial sites, etc. As with other Web sites lacking medical authority, accurate information is not guaranteed, but the personal experiences chronicled on some of these pages may be especially helpful and comforting.

After conception, the odyssey of pregnancy typically begins with a pregnancy test. To learn about these devices, along with ovulation predictor kits and other home medical tests, consult the Home Diagnostic Kits Home Page <http://kerouac.pharm.uky.edu/HomeTest/KitsHP.html>. Written by a chemist at the University of Kentucky, this site includes descriptions of how the tests work, instructions for use, comparison of different brands, answers to common questions, and more. Once the test comes back positive, a woman's next step is usually to locate a physician. AMA Physician Select Online

Doctor Finder <http://www.ama-assn.org/aps/amahg.html> includes information on over 650,000 MDs and DOs in the U.S. and can be searched by physician name or specialty and location. The ABMS Certified Doctor Home Page <http://www.certifieddoctor.org/> offers a certification verifier and certified doctor locator. Given a doctor's name and location, the verifier allows a user to verify the specialties in which the doctor is board certified. With the locator, a user can enter a location and a specialty and receive a list of certified doctors. However, the locator only includes doctors who have paid a fee to be listed; it does not include all board certified doctors. Women who prefer a midwife to an MD can go to Midwifesearch.Com <http://www.midwifesearch. com/> for a searchable directory of midwives and doulas. The midwife directory includes both certified nurse midwives and lay or direct-entry midwives. Other childbirth professionals can also be found via the Web. Doulas of North America (DONA) <http://www.dona.com/> offers a directory of DONA-certified doulas, arranged by location, with the majority from the U.S. and Canada, plus general information about doulas. ICEA Certified Childbirth Educators <http://www.icea.org/icces.htm> features a directory of childbirth educators in the U.S., Canada, and other countries.

Once a woman has found a health care professional and begun prenatal care, she will soon face prenatal tests. A visit to Ultrasound and Other Prenatal Diagnostic Tests <http://www.stanford.edu/~holbrook/> will explain ultrasound, amniocentesis, chorionic villus sampling, and alpha fetoprotein screening in consumer-friendly language. If one of these tests should reveal a multiple pregnancy, a visit to Parents of Twins/Multiples <http://www.twinslist. org/> is in order. This site, the home page for the Twins and Supertwins e-mail list, includes links to the list's FAQ's, an extensive archive of tips from parents on surviving pregnancy (including birth stories), preparing for multiples, parenting multiples, coping with the difficult early postpartum weeks, managing multiples as toddlers, and traveling with multiples.

Information on uncomplicated pregnancy is readily available from many of the sites already listed. The large, general sites, however, often do not cover pregnancy complications thoroughly. The following sites will help fill this gap. High Risk Pregnancy Resources <http://members.aol.com/MarAim/ bedrest.htm> is a list of books, Web sites, audio tapes, and organizations relating to a variety of pregnancy complications and issues, including bed rest, intrauterine growth retardation, preterm labor, and pregnancy loss. Some of the entries are annotated. The Sidelines National Support Network <http://www.sidelines.org/> provides news, e-mail support, articles on preterm labor and bed rest, a directory of state Sidelines chapters, a directory of organizations and other resources, weekly chats, and more. This is a wonderful resource for any woman coping with bed rest and/or preterm labor. Motherisk <http://motherisk.org/>, a collaborative project of Toronto's Hospital for Sick Children and the Society of Obstetricians and Gynaecologists of Canada (SOGC), includes detailed, and hard to find, information on cancer in pregnancy, plus information on several other conditions. The site can be difficult to navigate, and some material is written for physicians rather than consumers, but for someone facing one of the conditions documented here, the information will be worth the trouble it takes to get it. Finally, HIV-positive women will want to visit Pregnancy and HIV: What Women and Doctors Need to Know <http://www.hcfa.gov/hiv/default.htm>, a very useful site from the U.S. Health Care Financing Administration. It covers many issues related to HIV and pregnancy, such as HIV testing, reducing the risk of HIV transmission from mother to baby, and Medicaid coverage. The information is written in clear, plain English with helpful illustrations, where appropriate.

Once a woman has survived pregnancy and childbirth, the joys and challenges of the postpartum period begin. For women choosing to breastfeed, the Web offers a plethora of sites with helpful information. While most of the medical, pregnancy, and parenting sites offer some information, more detailed information can be obtained from La Leche League International <http://www.lalecheleague.org/>. The site is a complete source of breastfeeding information, with a wonderful collection of articles describing first-person experiences with breastfeeding, including nursing preemies, multiples, and adopted babies. Also available is a detailed FAQ covering many aspects of breastfeeding and a directory of La Leche League leaders in the U.S. and around the world, with links to local groups. Some information is available in Spanish, Italian, German, Chinese, Dutch, Flemish, French, Hebrew, and Japanese. More clinical information on breastfeeding is available as a Case Western Reserve University online course entitled Breastfeeding Basics <http://www.breastfeedingbasics.org/>, designed to teach the fundamentals of breastfeeding to health professionals. Despite its intended audience, course material is clear and understandable to educated consumers. The course covers anatomy and physiology, growth and development, breastfeeding around the world, the breastfeeding couple, the term infant with problems, and breast milk and drugs. Some sections include patient information handouts in PDF format. For more personal experiences, and sometimes aggressive advocacy of breastfeeding, one can visit the WWW Breastfeeding Ring <http://users.aol.com/kristachan/bfring/bfring.htm>, a Web ring of 156 mostly noncommercial sites, many of which are personal pages. As with other Web rings, the quality varies from site to site, and medical accuracy cannot be guaranteed.

It is estimated that postpartum depression affects women in 3% to 20% of all births, while the more common "baby blues" affect women in 50% to 80% of all births.[1] Information for these women and their partners is available from Postpartum Depression: A Social Support Network, Information

Center and Research Guide <http://www.chss.iup.edu/postpartum/>, a joint project of Postpartum Support International and the Indiana University of Pennsylvania Department of Anthropology. The site offers an excellent introduction to postpartum illnesses, a directory by state of support groups in the U.S., an article for new mothers on assessing and building social support, material on coping with anxiety, chat rooms, and more.

PEDIATRICS

Unlike computers, babies do not come with owner's manuals, and the closest thing to technical support is usually the pediatrician's office. Fortunately, there are a number of good Web resources that can help new parents learn about normal growth and development, common childhood illnesses and injuries, and more serious conditions. Information about normal development and common concerns and problems is also available from the general medical and parenting sites described in the first two sections of this article.

General Pediatric Sites

The American Academy of Pediatrics <http://www.aap.org/> maintains a site that includes news, recent research, and publications, plus a "You and Your Family" section with material for consumers. The AAP's public education brochures are available here, with information on a variety of medical issues and conditions, plus social and safety issues (e.g., violence, television). Also available are injury prevention tip sheets, information on Sudden Infant Death Syndrome, a directory of pediatric Internet sites arranged by category, and more. Several pediatricians maintain virtual offices on the Web, answering questions and publishing articles on pediatric topics. One of the best is Dr. Greene's House Calls <http://www.drgreene.com/>, by Dr. Alan Greene, a board certified pediatrician and author of several child care books. The questions he receives, and his answers, can be browsed by topic or searched by keyword. His answers are detailed, yet easy to understand and reflect current practice. Another good pediatrician site is Kids Doctor <http://www.kidsdoctor. com/>, maintained by Dr. Lewis A. Coffin, III. Dr. Coffin's articles cover a variety of topics related to illness, injury, nutrition, and allergies, and are searchable.

Though written for doctors and medical students, PEDBASE <http:// www.icondata.com/health/pedbase/index.htm> is useful for consumers because of its comprehensiveness. It documents over 550 childhood illnesses and conditions, including many lesser-known conditions that are not covered well elsewhere. The site is very plain, just an alphabetical list of conditions, but the list can be searched using a Web browser's "Find" function. Information on each condition is taken from at least three sources, including Nelson's *Textbook of Pediatrics,* the *Birth Defects Encyclopedia,* and various journal articles, review articles, and textbooks. The entries may be difficult for the layperson to understand, but for rarer conditions, PEDBASE may be one of only a few available sources.

Two specialized sites may also be of interest. Zero to Three: National Center for Infants, Toddlers, and Families <http://www.zerotothree.org/>, a nonprofit organization founded by developmental experts, offers a wealth of information on cognitive, psychological, and social development of babies and toddlers. Many of the articles are long and written at a high reading level, but they provide information not readily available elsewhere on the Web. The Immunization Action Coalition <http://www.immunize.org/>, a nonprofit organization established to boost immunization rates, offers a collection of patient education handouts on immunization-related topics, in PDF and HTML, some in Spanish. The site also includes a collection of patient education handouts, mostly on hepatitis B, in many languages, including Cambodian, Vietnamese, Russian, Somalian, and Korean, in PDF format.

To locate additional sites related to pediatrics, one can visit Pediatric Points of Interest <http://www.med.jhu.edu/peds/neonatology/poi.html>, a directory of pediatric Internet links from the Department of Pediatrics and the Residency Program at Johns Hopkins University. Some of the material is geared for physicians, but the site includes a parenting section and patient education links and covers a wide range of topics. Another option is the SLACK Pediatric Internet Directory <http://www.slackinc.com/child/pednet-x.htm>, which includes a variety of links for physicians and parents, including links to e-mail lists, newsgroups, and electronic newsletters.

Special Problems

When an infant has health problems, parents feel a range of emotions, from fear and anger to frustration at a situation they cannot control. While no Web site can make a child well, several sites can provide information to help parents understand a baby's condition and locate sources of help and support.

Ten percent of babies are born prematurely, before thirty-seven weeks gestation.[2] The University of Wisconsin and Meriter Hospital in Madison, Wisconsin, have created a great site, entitled For Parents of Preemies: Answers to Commonly-Asked Questions <http://www2.medsch.wisc.edu/childrenshosp/parents_of_preemies.index.html>, which includes information on chances for survival and disability, an introduction to the Neonatal Intensive Care Unit, handouts on common problems and diseases in preemies, a list of commonly-prescribed drugs, and more. The site is also available in Spanish. A sister site, My Sick Newborn: Answers to Commonly-Asked Questions <http://

www2.medsch.wisc.edu/childrenshosp/sicknewborn/t-index.html>, provides similar information for parents of a full-term baby with problems. It includes information on common abnormalities in development, problems and diseases, common questions after discharge, and much more. It covers many neonatal problems, including less common ones not mentioned at most parenting and pediatrics sites. For parents of preemies, additional information, along with lots of personal stories and words of encouragement, can be found in the Preemie Ring <http://members.aol.com/liznick1/preemiering.htm>, a collection of 227 noncommercial sites, most of which are personal home pages chronicling the lives and development of premature babies. Reliable medical information is not guaranteed, but seeing pictures and reading the story of a healthy child who was born at twenty-four weeks can provide hope for frightened parents.

For information on birth defects, the March of Dimes Fact Sheets <http://www.modimes.org/HealthLibrary2/Factsheets/Default.htm> are an excellent source. Fact sheets are available in the following categories: Drugs and Alcohol During Pregnancy, Birth Defects and Genetics, Infections/Diseases During Pregnancy, Newborn Information, Polio, Pregnancy, Pregnancy Loss and Other Concerns, and Prenatal Screenings. The pages are written in plain English but include many details and, unlike most fact sheets on the Web, include a list of references to the medical literature. Another source of information for genetic diseases is Rare Genetic Diseases in Children <http://mcrcr2.med.nyu.edu/murphp01/homenew.htm>, maintained under the aegis of the New York University Medical Center. It includes message forums, links organized by category (advocacy, disabilities, disease Web sites, support, etc.), a sourcebook on lysosomal storage diseases, and a layman's guide to genetic inheritance, the latter two items in both French and English. Parents of a child with any disability will also want to visit Internet Resources for Special Children <http://www.irsc.org/>, a site developed by the parent of a disabled child. It includes links organized by type of disability, with some annotations.

It seems incongruous to include death in a discussion of pregnancy and birth, but the unfortunate reality is that not every pregnancy results in a living child. For parents grieving the loss of a pregnancy or the death of a child, the Web offers both information and support. The Compassionate Friends <http://www.compassionatefriends.org/> is a national, nonprofit, self-help organization for families grieving the death of a child. Their site includes a collection of brochures for grieving families and their friends, links to local Compassionate Friends chapters, and lots more. M.I.S.S.: Mothers in Sympathy and Support <http://www.misschildren.org/> is a nonprofit organization providing emergency support to parents after the death of a baby. The site includes articles, poems, and newsletters on coping with grief and other issues related to the loss of a child. The SIDS Network: Sudden Infant Death Syndrome and Other Infant Death Information Web Site <http://sids-network. org/> includes information on preventing SIDS in fourteen languages, grief, and sibling grief, plus brochures, personal experiences, chats, discussion boards, and FAQ's on several topics related to SIDS, including smoking, vaccinations, and apnea and monitoring. Finally Usenet includes soc. support.pregnancy.loss, a newsgroup that serves as an online support group for pregnancy loss. The newsgroup's charter and FAQ's are available at <http://web.co.nz/%7ekatef/sspl/>.

While the Internet should not and cannot take the place of competent, compassionate medical care, the material found on the sites described here can help parents become informed participants in their own and their child's medical care and find reassurance that they are not alone in their experiences, trials, and tribulations. Parenthood is a wild, wonderful adventure, made a bit more enjoyable with the confidence born of sound knowledge and information.

NOTES

1. Kruckman, Laurence, and Smith, Susan. "An Introduction to Postpartum Illness." [Web page] Indiana University of Pennsylvania, 1998. Available: <http://www.chss.iup.edu/postpartum/preface.html#C>. Accessed: January 3, 2000.

2. Kimsey, Bobbi. "Born Too Early: Preterm Labor and Birth." [Web page] Medical University of South Carolina, June 1996. Available: <http://www.musc.edu/nursing/faculty/midwifery/pregncy/ptl.htm>. Accessed: January 3, 2000.

A Guide to Web Resources
for Caregivers

Deborah Sobczak

ABSTRACT. Family caregiving is one of the nation's valuable assets. Caregivers dedicate their time and energy to taking care of family members who can no longer care for themselves. Caregivers feel isolated and overburdened from balancing caregiving responsibilities, careers, marriages, and children. The profile of a caregiver is a middle-aged working woman with children. The issues of caregiving are receiving national attention as more Americans are living longer. One of the challenges in the twenty-first century is the senior boom. Caregiving has been part of the family experience, but as a nation we must address the issues of long-term care. *[Article copies available for a fee from The Haworth Document Delivery Service: 1-800-342-9678. E-mail address: <getinfo@haworthpressinc.com> Website: <http://www.HaworthPress.com>]*

KEYWORDS. Caregiving, caregivers, family caregiver, long-term care, chronic illness, aging, Internet

INTRODUCTION

Every day in this nation, men and women give generously and without compensation their time and energy to taking care of family members who can no longer take care of themselves. These ordinary heroes in our society

Deborah Sobczak, MSLS, AHIP, is Information Services Librarian, Helen L. DeRoy Medical Library, Providence Hospital & Medical Centers, P.O. Box 2043, 16001 West Nine Mile Road, Southfield, MI 48037.

[Haworth co-indexing entry note]: "A Guide to Web Resources for Caregivers." Sobczak, Deborah. Co-published simultaneously in *Health Care on the Internet* (The Haworth Press, Inc.) Vol. 4, No. 2/3, 2000, pp. 113-120; and: *Women's Health on the Internet* (ed: M. Sandra Wood, and Janet M. Coggan) The Haworth Press, Inc., 2000, pp. 113-120. Single or multiple copies of this article are available for a fee from The Haworth Document Delivery Service [1-800-342-9678, 9:00 a.m. - 5:00 p.m. (EST). E-mail address: getinfo@haworthpressinc.com].

balance work, marriage, and children to improve the well-being and quality of life of an elderly or disabled parent, grandparent, relative, or friend. Family caregiving is one of the nation's valuable assets and saves the health care system billions of dollars a year. Caregivers must rearrange work schedules, work fewer hours, and take unpaid leaves of absence to meet the demands of caregiving. A caregiver makes significant work sacrifices and may not achieve greater career goals if he/she is a primary or secondary caregiver. A study conducted by the National Center for Women and Aging at Brandeis University and the National Alliance for Caregivers estimated the average loss in wages, pension, and Social Security benefits over a caregiver's lifetime is $659,139.[1] If these informal caregivers were replaced by paid home care staff, the cost to the nation's health care system would be a staggering $45 to $75 billion per year.[2] Often, caregivers feel isolated and overburdened from balancing caregiving responsibilities. Caregivers are affected by both negative and positive emotions, from frustration and depression to enlightenment that they found the inner strength to meet the challenges of caring.

BACKGROUND

Advances in medical science and technology have allowed more Americans over 60 years of age the opportunity to add years to their retirement to enjoy leisure activities, second careers, or volunteerism. By 2030, 76 million baby boomers will join the ranks of the retired; doubling the number today.[3] By the middle of the next century, the average American will live to an age of 82–that's six years longer than the average life expectancy today.[4] However, a growing number of aging Americans, who are in declining health and economic status, need supportive services from a caregiver. An estimated one out of every four households (23% or 22.4 million) in the United States is providing essential, long-term care to an elderly family member.[5] One of the central challenges in the twenty-first century is the senior citizen boom.

WHO ARE THE CAREGIVERS?

The profile of a typical caregiver is a middle-aged woman, working full time, married, and a mother with children not old enough to be independent. In other words, a female baby boomer who's part of the "Sandwich Generation," caught between caring for her children and aging parents. The costs and challenges of providing care to a loved one with a long-term illness can put a staggering emotional and financial strain upon the family. Almost all of home care is provided by an informal caregiver; an unpaid individual who provides care. Only 14% of home care is provided by paid caregivers.[6]

RAISING AWARENESS OF CAREGIVING

In 1993, President Clinton signed the Family and Medical Leave Act (FMLA) into law, providing benefits for workers to take up to twelve weeks unpaid leave to care for a new baby or ailing family member without jeopardizing their job. Every year, the President designates the week of Thanksgiving as "National Family Caregivers Week," to nationally recognize the daily contributions of family caregivers in communities throughout the nation. Efforts such as these can be helpful to increase the awareness of family caregiving. In September 1998, Rosalynn Carter brought more attention to caregiving issues when she addressed the Senate Special Committee on Aging. She candidly shared her experiences as a caregiver, first to her own family, and then to her husband's relatives. In her testimony, she credited a colleague for describing how prevalent caregiving is. "There are four kinds of people in this world: those who have been caregivers; those who currently are caregivers; those who will be caregivers; and those who will need caregivers."[7] Her work at the Rosalynn Carter Institute <http://rci.gsw.edu> on the campus of Georgia Southwestern State University focuses on the need for federal, state, and local governments to offer increased support for caregivers. Mrs. Carter advocated expanding Medicare to cover respite care costs. Respite care programs can offer much needed time off for families from the overwhelming tasks of caregiving. A substitute caregiver visits the home, sometimes everyday, to give the primary caregiver some necessary relief.

In January 1999, President Clinton announced four proposals totaling $6.2 billion (over five years) to help families fulfill their obligations to loved ones needing long-term care. The first initiative will provide an unprecedented $1,000 tax credit that compensates, for formal or informal costs, Americans of all ages with long-term care needs or the family caregivers who support them. The second proposal creates the establishment of the National Family Caregiver Support Program (NFCSP) that will enable families to receive community-based counseling and support. The third proposal will educate Medicare beneficiaries about their long-term care choices. Medicare, as well as most private insurers, does not cover most kinds of long-term care, so it is important that beneficiaries understand their options. The fourth proposal is to use the federal government, the nation's largest employer, as an example for the industry in providing private long-term care insurance to federal employees. By November of 1999, 40 federal bills related to family caregiving were introduced in Congress. Despite White House and the U.S. Department of Health and Human Services support, not one bill has come to fruition.

WHO NEEDS CAREGIVERS?

In 1995, 100 million Americans suffered with one or more chronic health conditions.[8] About 12.8 million Americans need assistance with daily activities such as bathing, dressing, and eating.[9] Caregivers spend on average 22 hours a week helping their elderly relative with the grocery shopping, transportation to doctor appointments, and meal preparation.[10] Women constitute 73% of all caregivers, and 79% of them give constant care.[11] The level of caregiving responsibilities varies by the scope and severity of the chronic illness. An estimated four million Americans are suffering from Alzheimer's Disease; a devastating, irreversible cause of dementia in older adults.[12] The aging population could mean an epidemic of patients with chronic illnesses during the next 50 years.

DEMANDS OF CAREGIVING

The demands of being a caregiver may lead to chronic stress, depression, or an early death. A recent study in the December 15, 1999 issue of *JAMA* indicated that an elderly spouse who experienced strain related to providing support to their disabled spouse is more likely to have a higher mortality rate than non-caregivers.[13] No significant increase in mortality was reported among spouses who did not have economic, physical, or emotional distress associated with caregiving. The study suggested that being a caregiver has health risks; especially among elderly spouses. Research studies are recognizing the impact of caregiving on the health of the caregiver.

WEB RESOURCES FOR CAREGIVERS

The following list of suggested links originates from government agencies, well-respected non-profit agencies, and private organizations. This is by no means a complete representation of Web resources available to date. Brief annotations are meant to describe each link's attributes and features in order to provide searchers with pertinent information before they visit a particular site. The following collection of useful Web resources can assist health sciences librarians in searching for caregiving information on the Web.

Administration on Aging
<http://www.aoa.dhhs.gov>

The navigational tools on this site point to an abundance of information for older persons and their families (ElderPage), practitioners and other health professionals, and researchers and students. ElderPage contains "ElderAction," a series of online pamphlets on major topics such as retirement living, fitness, transportation, and volunteer opportunities. The AoA Fact sheets and the National Institute of Aging Age Pages contain a variety of links related to aging and caregiving issues from elder abuse to home care. Other links to online publications include "Talking to Your Doctor," "Retirement and Financial Planning," "Prescription Medicines and You," and "Guide to Choosing a Nursing Home." The AoA site features the Eldercare Locator, a service for finding directory assistance to help older persons and caregivers locate local support resources.

AgeNet.com
<http://www.agenet.com>

Caregivers.com
<http://www.caregivers.com>

AgeNet is a site offering a vast array of topical issues for aging adults and their families. Membership to AgeNet is free and provides access to chatrooms and message board areas. The table of contents includes links to geriatric health, geriatric drugs, legal, insurance, financial, helpful products, and caregiver support. Visitors to the site interested in housing and living alternatives have the opportunity to conduct a national, state, and local search for nursing homes, assisted living facilities, home health agencies, and retirement communities. Caregivers.com is sponsored by AgeNet and focuses primarily on caregiving issues.

Association of American Retired Persons (AARP)
<http://www.aarp.org/caregive/home.html>

Since 1958, the AARP is the nation's largest organization for aging Americans. Anyone age 50 or older, either retired or still employed, can join. The AARP is host to a weekly online caregiver support group on America Online. A wealth of information is provided about the many services, support systems, and publications available to caregivers.

Caregiver Survival Resources
<http://www.caregiver911.com>

Jim and Merlene Sherman developed this site to aggregate information about their workshops, books, and caregiving experiences. There are five categories ranging from books, resources, ask-the-expert, and community bulletin boards.

Caregiving Online
<http://www.caregiving.com>

Caregiving Online is an offshoot of the monthly newsletter, *Caregiving,* edited and published by Denise Brown. The "Caregiver of the Year" award is an annual event recognizing a special family caregiver. Each December, the profile of the winner appears in the issue and on the site.

Center of Family Caregivers
<http://www.familycaregivers.org>

Denise Brown is the executive director of this virtual organization. Their mission is to help family caregivers understand their roles by sponsoring workshops and developing educational materials. The first project was a kit to help caregivers with stress reduction.

Eldercare
<http://www.eldercare.com>

Eldercare is a comprehensive, free resource to meet the needs of caregivers providing local or long-distance care. The site offers access to a host of checklists, brochures, directories, and articles, and publications. The Finding Care Services directory is searchable by state, metropolitan area, and city.

Family Caregiver Alliance
<http://www.caregiver.org>

The Family Caregiver Alliance, a non-profit, community-based service, was formed in 1977 to address the needs of families providing long-term care to the elderly. The Clearinghouse has a full listing of FCA publications, statistics on long-term care, recommended readings, and information on specific diagnoses and disorders.

National Alliance for Caregiving
<http://www.caregiving.org>

This is a non-profit organization with several partner organizations and affiliates. The Alliance has completed several studies on caregiving in the U.S. available in pdf format. The Director's Corner lists book and video suggestions applicable to a wide range of caregivers. Some links, such as the Care Wizard and the Study Abstracts, were still under construction.

National Family Caregivers Association
<http://www.nfcacares.org>

The aim of the NFCA is to promote the well-being of family caregivers through education, advocacy, and public awareness. An important feature on the site was Caregiving Tips. These guidelines may help steer a caregiver in the right direction or give some much-needed validation.

Well Spouse Foundation
<http://www.wellspouse.org>

The Well Spouse Foundation is a newcomer to the Web. It's a national, non-profit organization providing support for spouses and partners of the chronically ill or disabled. A virtual community of caregivers is provided as a service for members. The WSF publishes pamphlets that address unique experiences and issues of caregivers. *Mainstay* is the bimonthly newsletter of the organization; however, current articles were not available on the site.

CONCLUSION

Caregiving has been embedded in the family experience, history, and values for centuries. Modern medicine has enabled more Americans to live longer lives and to live with terminal illnesses that one day will cause their deaths. Caregivers' efforts enhance the quality of life of older adults but a majority of today's caregivers have made no plans for their own long-term needs. As a nation, we need to look ahead and address these issues of long-term care. One day in the future, caregiving may affect every American family.

NOTES

1. Metropolitan Life Insurance Company. "Americans Pay a Staggering Price in Lost Wages and Other Costs to Care for Elderly Relatives and Friends According to MetLife Study." December 1, 1999. Available: <http://heller.brandeis.edu/national/metpress.htm>. Accessed: December 7, 1999.

2. Administration on Aging. "Family Caregiver Fact Sheet." Available: <http://www.aoa.gov/May99/caregiver.html>. Accessed: December 7, 1999.

3. Office of the Press Secretary. "Remarks by the President on Long-Term Health Care Initiative." 4 January 1999. http://www.pub.whitehouse.gov/WH/Publications/html/Publications.html (10 December 1999).

4. Office of the Press Secretary. "Remarks by the President on Long-Term Health Care Initiative." January 4,1999. Available: <http://www.pub.whitehouse.gov/WH/Publications/html/Publications.html>. Accessed: December 10, 1999.

5. Family Caregiver Alliance. "Fact Sheet: Selected Caregiver Statistics." March 1998. Available: <http://www.caregiver.org/factsheets/caregiver_statsC.html>. Accessed: December 7, 1999.

6. Family Caregiver Alliance. "Fact Sheet: Selected Caregiver Statistics." March 1998. Available: <http://www.caregiver.org/factsheets/caregiver_statsC. html>. Accessed: December 7, 1999.

7. United States Senate. "Rosalynn Carter's Testimony. Senate Special Committee on Aging. Hearing on Caregiving Issues." September 10, 1998. Available: <http://www.senate.gov/~aging/hr24rc.htm>. Accessed: December 7, 1999.

8. Hoffman, C.; Rice D.A.; and Sun, H.Y. "Persons With Chronic Conditions: Their Prevalence and Costs." *JAMA* 276(November 13, 1996): 1473-9.

9. Family Caregiver Alliance. "Fact Sheet: Selected Caregiver Statistics." March 1998. Available: <http://www.caregiver.org/factsheets/caregiver_statsC.html>. Accessed: December 7, 1999.

10. National Alliance for Caregiving. "The Caregiving Book: Baby Boomer Women Giving Care." September 1998. Available: <http://www.caregiving.org/content/reports/babyboomer.pdf>. Accessed: December 7, 1999.

11. National Alliance for Caregiving. "Family Caregiving in the U.S. Findings from a National Survey." June 1997. Available: <http://www.caregiving.org/content/reports/ finalreport.pdf>. Accessed: December 7, 1999.

12. Administration on Aging. "Alzheimer's Disease; Administration on Aging Fact 13. Sheet." Available: <http://www.aoa.dhhs.gov/factsheets/alz.html> Accessed: December 7, 1999.

13. Schulz, Richard. "Caregiving as a Risk Factor for Mortality." *JAMA* 282(December 15, 1999):2215-9.

Charting the Web to Navigate "The Change": Consumer Resources for Menopause on the Web

Valerie A. Gross
Patrice Hall

ABSTRACT. As women approach menopause, many desire reliable and easy-to-understand information about this next phase of their life. The Web has become one resource to which they can turn. This article is a guide to some of the better sites. It includes information on Web directories and search engines, general health sites, and those sites devoted specifically to women's health and menopause. *[Article copies available for a fee from The Haworth Document Delivery Service: 1-800-342-9678. E-mail address: <getinfo@haworthpressinc.com> Website: <http://www.HaworthPress.com>]*

KEYWORDS. Menopause, women's health, World Wide Web

BACKGROUND

During the next decade, 40 million women will turn 50.[1] This latest permutation of the baby boom population has already had far reaching effects in

Valerie A. Gross, MLS (vgrass@psghs.edu), is Women's Resource Center Librarian, Geisinger Medical Center, 100 N. Academy Ave., Danville, PA 17822-2020. Patrice Hall, MLS (chil@psghs.edu), is Librarian, Community Health Information Library, The Milton S. Hershey Medical Center, P.O. Box 850 HS-07, Hershey, PA 17033-0850.

[Haworth co-indexing entry note]: "Charting the Web to Navigate 'The Change': Consumer Resources for Menopause on the Web." Gross, Valerie A., and Patrice Hall. Co-published simultaneously in *Health Care on the Internet* (The Haworth Press, Inc.) Vol. 4, No. 2/3, 2000, pp. 121-131; and: *Women's Health on the Internet* (ed: M. Sandra Wood, and Janet M. Coggan) The Haworth Press, Inc., 2000, pp. 121-131. Single or multiple copies of this article are available for a fee from The Haworth Document Delivery Service [1-800-342-9678, 9:00 a.m. - 5:00 p.m. (EST). E-mail address: getinfo@haworthpressinc.com].

© 2000 by The Haworth Press, Inc. All rights reserved.

121

many areas of our society but particularly in the health care field. In the past few years, there has been an increase in the medical research concerning the health of midlife women with an accompanying explosion of health information.[2] However, controversy and contradictions reign in this area. Should menopause be considered a natural process to be endured or a disease that can be treated? No longer a taboo subject, women are talking openly about menopause and actively seeking information to help them understand and make decisions about this life change. The World Wide Web is one resource that women can turn to for a better understanding of this life event.

INTRODUCTION

The purpose of this article is to identify and describe useful Web sites that contain information about menopause or index such sites. This list is selective rather than exhaustive. However, it does endeavor to be inclusive by including sites on both sides of the natural process/disease process controversy. Web users need to remember that it is up to the individual to evaluate the credibility and usefulness of any information posted on the Web.

WEB RESOURCES

Medical Web Search Engines and Directories

One way to access the resources on the Web is to do a subject search using either a search engine or a directory. Although it is possible to use the general search sites, such as Google or Yahoo!, to locate menopause information, using medical search sites is preferable as these eliminate some of the "noise" of false hits and unreliable sites that can turn up using general search tools.

healthfinder®
<http://www.healthfinder.gov>

Developed by the U.S. Department of Health and Human Services, this site is uncluttered and easy to search by either keyword searching or browsing. Indexed sites include government agencies; national voluntary organizations, nonprofit and professional organizations; educational institutions; and libraries. Each resource has a details link that gives a description of the resource and its sponsoring organization. The details link for organizations includes a description of the organization and contact information.

For information on menopause, the user should access both the alphabetical index and the search engine, as there are some differences in the search results.

MEDLINE*plus*
<http://www.nlm.nih.gov/medlineplus/>

The National Library of Medicine maintains this government site, MEDLINE*plus*. Indexed sites have an emphasis on information available from NLM and its parent organization, the National Institutes of Health, such as the many full-text publications produced by the NIH institutes.

Geared more toward browsing, the site offers a variety of access points. Users can select a topic from either an alphabetical list or from a list of categories. There are also links to tools such as dictionaries, health care providers and hospital locators, lists of organizations, clearinghouses, libraries and other health databases. In addition to the browsing features, there is a search engine. Search results are annotated, and the more relevant matches are rated with stars.

Another useful feature of this site is the preformulated MEDLINE searches for each topic. This feature enables users to easily access citations from the latest medical journals. For example, under the topic of menopause, there are four searches available: general, premenopause, premature menopause, and therapy for menopausal symptoms. Choosing one of these searches gives the consumer a list of citations. One thing to keep in mind when using the MEDLINE searches is that, although there is a link to order articles, the consumer has to have an account established with a library. Usually this is a service only available to the staff of medical or academic libraries. Consumers need to contact their local consumer health or public library about obtaining the articles through interlibrary loan services.

Doctor's Guide to the Internet
<http://pslgroup.com>

Designed by P\S\L Consulting Group Inc. as an indexing service to Web resources for physicians, this directory can also be useful to consumers. Its home page is divided into three major sections. The first section provides links to sites of interest to health professionals, including news services, drug information, and CME and conference information. The second section is arranged by health topic and includes sites that have consumer health information. Menopause is one of the topics covered. The third section contains links to travel sites.

To locate menopause information the user should choose the menopause link in the second section. This will take the user to an index of abstracts and summaries of medical journal articles and conference reports, links to menopause articles from other Web sites, links to support groups and discussions lists and other consumer health Web sites. Although this is not the best site to search for basic menopause information, its strength is its links to current news, particularly about the latest drug developments.

General Health Information Sites

General health information sites provide original content from the sponsors and content partners of the site. Many general sites include interactive features like tests and quizzes, message boards, and chat rooms. These sites also include links to other reliable Web pages.

Mayo Clinic Health Oasis
<http://www.mayohealth.org/>

The Mayo Foundation for Medical Education and Research is the sponsor for this site of prolific resources. The site contains a wealth of health information for consumers. Health Oasis offers a feature article of the day, access to the Oasis Library, the ability to send e-mail to a Mayo Clinic physician, the ability to search the site for specific information, ten centers of health information, and much more. The ten centers of health information are: "Allergy & Asthma," "Alzheimer's," "Cancer," "Children's," "Digestive," "Heart," "Medicine," "Men's," "Nutrition," and "Women's." Going to the "Women's Center" is the way to find information about menopause that is grouped into categories. To find the grouped categories, the consumer must enter the section entitled "Reference Articles." The section on menopause provides information about perimenopause, hormone replacement therapy, incontinence, exercise, sexuality during midlife, and weight gain during menopause. There is also a very informative section about osteoporosis.

Another way to obtain information about menopause is to perform a general search. By doing this, consumers will find information from a variety of sources. The sources range from answers to "Ask the Mayo Physician," to articles from the Mayo Clinic Health Letter. It requires some time to peruse the results from a general search, so if time is of the essence, going directly to the "Women's Center" to find menopause information is recommended.

drkoop.com
<http://drkoop.com>

The chairman of the board for this for-profit venture is the former U.S. Surgeon General, Dr. C. Everett Koop. A number of institutions contribute to the content of this site: The American Council on Science and Health (ACSH), The Cleveland Clinic Foundation, Dartmouth Medical School, Lifescape.com, Multum Information Services, Inc., Screaming Media, and World Book, Inc.

The first-time user of this site may be a bit overwhelmed. There is so much good information it is hard to know where to begin. Consumers can find sections about "Health News," "Family Health," "Health Resources," "Health & Wellness," "Community," and "Conditions & Concerns." There are "Disease Centers," "Featured Sections," and an excellent drug interaction database. For the purpose of finding menopause information, consumers may click on the menopause section in the box labeled "Browse through more than 60 health topics." There are a number of ways to find menopause information on this site, but browsing through the " . . . more than 60 health topics" seems to be the easiest way. Once the consumer has arrived at the menopause area, he or she is presented with more choices. The two major choices are a link to the definition of menopause and a link to a menopause library. In the library, consumers will find information about hormone replacement therapy, heart disease, osteoporosis, sexuality, psychological aspects, hot flashes, cycle changes, urogenital tract information as well as information about health care providers. The library is an excellent source for women just beginning their quest for menopause information.

HealthAtoZ
<http://www.healthatoz.com>

Medical Network, Inc. (MNI) sponsors this site. The Medical Advisory Board consists of health professionals and physicians from the Philadelphia College of Pharmacy, Harvard Medical School, Georgetown Providence University Hospital, Pennsylvania Hospital at the University of Pennsylvania, University of Maryland Hospital, and many other in reputable institutions.

Similar to previously reviewed sites, HealthAtoZ has an abundance of health information for the consumer. To find menopause information, the consumer may click on menopause in the "health topics AtoZ" box. Under the menopause section here, consumers will find brief information about perimenopause, menopause, and hormone replacement therapy. Signing up for the free membership gives the consumer access to even more features, including "Ask the Librarian" and "Ask the Experts" pages and the "Community" features, which provide special presentations from health professionals.

This site also includes a Web search engine. Typing in menopause brings up links to over 200 Web documents, all of which have been reviewed and rated by HealthAtoZ's staff of health care professionals.

The Foundation for Better Health Care
<http://fbhc.org/>

The goal of this nonprofit corporation is to provide information and education to patients and professionals. This site provides nine consumer health topic areas, one of which is menopause. The site is uncluttered and the consumer health sites are easy to locate. It contains information about perimenopause, hormone replacement therapy, osteoporosis, and heart disease, and answers to frequently asked questions. This site would be a good place to start looking for general menopause information.

HeliosHealth.com
<http://helioshealth.com/>

HeliosHealth has information centers, features, and much more. The Medical Advisory Board has as its members health care professionals from Vanderbilt University School of Medicine, University of Miami School of Medicine, American College of Obstetricians and Gynecologists, private practice, and other national associations.

The menopause section is easy to locate, under the Information Center area. This is an excellent site for basic and in-depth menopause information. It includes information about current menopause news items, testing for menopause, the role of hormones, hormone replacement therapy, phytoestrogens, vitamins, hot flashes, mood swings, weight gain, stress, heart disease, osteoporosis, and Alzheimer's disease. A nice feature about this section is the link to the National Library of Medicine's database, MEDLINE. Using MEDLINE, consumers can search for more clinical information.

InteliHealth
<http://www.intelihealth.com>

Aetna U.S. Healthcare® and Johns Hopkins University and Health System sponsor this site. It is a busy site with animated graphics and frames, but the advertising banners are kept to a minimum. Although the home page is a bit distracting, the information is from a reliable source and has been reviewed by health care professionals from Johns Hopkins.

Menopause information can be accessed in a number of different ways. The user can browse the Women's Health section or click on "menopause" in "More Featured Areas." There is also a search engine for the site. Searching on menopause returns documents covering general menopause information, HRT, health complications, and midlife sexuality.

Medscape's Women's Health Page
<http://www.medscape.com/Home/Topics/WomensHealth/womenshealth.html>

Medscape is a for-profit Web site. The Editor in Chief is George D. Lundberg, former Editor of the *Journal of the American Medical Association* (*JAMA*). The editorial board consists of 20 experienced health care professionals. This site is for professional health care providers as well as consumers. Medscape has paired with the CBS Corporation to produce a consumer health site <http://cbs.healthwatch.com>, which, as of yet, has no specific information on menopause.

This page is for the more clinically minded consumer. Access to MEDLINE, TOXLINE, AIDSLINE, and the National Drug Data File is available. The site provides current research articles dealing with women's issues in fields such as obstetrics and gynecology, oncology, embryology, neurology, and cardiology. Some of the current research articles deal with aspects of menopause. A nice feature of the site is the "Risk Calculator For HRT." To use this feature and others, the consumer must register, but registration is free. Specific menopause information can be found by using Medscape's library. The menopause section gives the consumer numerous full-text clinical articles and continuing medical education from the past twelve months.

Menopause Matters ™
<http://www.menopausematters.com/menopausematters/index.htm>

Rhone-Poulenc Rorer, a global pharmaceutical company, sponsors this site. Completely dedicated to menopause information, their site covers general information about menopause, treatment options, mind/body, and support group resources. The information is from reliable sources such as the American College of Obstetricians and Gynecologists, and the *European Menopause Journal*. This is a good site with basic information.

JAMA Women's Health Information Center
<http://www.ama-assn.org/special/womh/womh.htm>

Ortho-McNeil Pharmaceutical and Advanced Care Products support the site. The American Medical Association generates much of the content and maintains the site. While the site is geared to the health care professional, consumers will find clinical articles about menopause. The site is uncluttered and contains headline news stories from professional journals and other sources, an STD center, a contraceptive center, and a library. The library contains current articles from journals such as *JAMA, Archives of Internal*

Medicine, BMJ, and *Journal of Clinical Oncology,* all of which may or may not address issues related to menopause. To find menopause-related articles, the consumer must perform a search. The results are ranked according to relevance. Articles are from the above-mentioned journals.

Menopause and Beyond
<http://www.oxford.net/~tishy/beyond.html>

This site is intended for those women who are on the threshold of menopause or who are postmenopausal. There is information related to menopause on cardiovascular disease, osteoporosis, breast concerns, pelvic organ disease, vaginal dryness, incontinence, hot flashes, and more. There is also a section on complementary medicine. Much of the information is not original content, but is taken from various reputable sources such as NIH, the American Heart Association, *British Medical Journal,* and *Lancet.* This site also addresses the psychosocial aspects of menopause, including mythology and poetry. One of the most interesting areas is the "Historical Perspectives" section, which contains medical literature about menopause written as early as 1869.

ThirdAge Women's Health Center
<http://www.thirdage.com/health/women/>

Mary Furlong, Ed.D., is the founder of ThirdAge Media, Inc. Partners include Barnesandnoble.com, Online Bookseller, Big Networks, Careguide, Caregiving Center Critical Path, and Online Digital Coupons and Incentives. Sponsors include Merrill Lynch, IBM, and Intel. This site is geared for older adults. It has weekly news updates, a diet and fitness center, newsletters, and information on menopause. There is a section where women may find out if they are perimenopausal, at risk for heart disease and osteoporosis, emotionally and sexually healthy, and if hormone replacement therapy is the right option for them. Also available is information about complementary therapy and myths about menopause. The information is brief, basic, and helpful.

North American Menopause Society
<http://www.menopause.org>

The North American Menopause Society is a nonprofit scientific organization whose members include professionals from the fields of medicine, nursing, sociology, psychology, nutrition, anthropology, epidemiology, and education. Their stated purpose is to promote the understanding of menopause, and thereby improve the health of women. Their Web site contains information for both the professional and the consumer.

The site is easy to navigate. Resources for the consumer include a section on basic menopause facts, frequently asked questions, information on natural remedies for menopause symptoms, and a 50-page booklet, "Menopause Guidebook," which provides a good overview of the physical and emotional changes that often accompany menopause.

National Women's Health Information Center
<http://www.4woman.gov>

The Office on Women's Health in the Department of Health and Human Services maintains this site. Its intended audience is both the consumer and health care professional. The information comes from federal agencies and reliable health organizations. It is searchable by either key word or by browsing the health topics section. Information on menopause can be located by either method. Also, the Frequently Asked Question section has two menopause entries. Both cover the same menopause information in a question and answer format, with one listed as "Easy to Read."

Women's Health Interactive: The Midlife Health Center
<http://www.womens-health.com/health_center/midlife/index.html>

This is a commercial site developed by Women's Health Interactive, a company that specializes in designing educational material for digital technologies. Affiliates and sponsors for this site include the National Headache Foundation, National Council on Women's Health, WomenWorks, Inc., Glaxo Wellcome, Inc., and Mothernature.com. The site is attractive and informative. Although it is a commercial site, it is free of advertising banners.

The site is arranged by centers that cover many aspects of women's health issues, one of which is the Midlife Health Center. This center includes interactive risk assessments on identification of perimenopause, severity of symptoms, health risks for osteoporosis, and heart disease. Topics covered in learning modules include long-term health risks, HRT, natural therapies, and sexuality. Other features include a national directory of women-centered service providers, frequently asked questions, support groups, and a personal action plan. The user must register to use some of these features.

ObGyn.Net
<http://www.obgyn.net>

This is a doctor-reviewed site with information for both the professional and the consumer on all aspects of women's health. It has articles written by subject specialists and provides over 4,500 links to other medical sites.

The home page, using an eye-catching graphic of planets, is divided into three main sections. One section is designed for health care professionals, one for the medical industry, and one for women. To find information on menopause, the user can click on the planet that is labeled "For Women and Patients." In this section, menopause is listed in both the "Featured Topic" frame and under the "Diseases and Conditions" bar. Both link to the "Menopause/Perimenopause" section. Here the user will find original articles written by the Ob/Gyn.Net editorial advisors and articles reprinted from medical journals and other Web sites. Other features include "Ask the Expert" message board, links to a number of "Doctor Finder" sites, a book list, and links to related sites on topics such as hysterectomy and osteoporosis.

Power Surge
<http://www.power-surge.com>

A commercial site founded by Alice Stamm in association with America Online and Thrive, this site would not be a place for finding basic, unbiased menopause information. It heavily promotes one soy product. However, it is a good site for those interested in joining a virtual community of women experiencing menopause. The main feature is its "Guest Conferences" with well-known authors and experts in the area of women's health. Past guests have included Dr. Susan Love, Lonnie Barbach, Gail Sheehy, Dr. Susan Lark, and Dr. Christiane Northrup. Other features include message boards, chat, and extensive links to other menopause and women's health sites.

Birthing the Crone
<http://www.birthingthecrone.com>

This site features menopause represented through the works of the artist, Helen Redman, who sees her art as "embracing menopause as a metaphor for transformation." It also presents an interesting discussion about the historical, philosophical, and societal aspects of menopause.

Related Sites

There are many other sites on the Web that cover health issues related to menopause, such as the increased risks of osteoporosis and cardiovascular disease. Some examples follow.

American Heart Association
<http://amhrt.org>

For those interested in the latest research on menopause and its relationship to heart disease, this site provides solid information.

National Osteoporosis Foundation
<http://www.nof.org>

This site contains information about the changes that can occur during and following menopause, and how these changes can affect a woman's bone health.

American Cancer Society
<http://www.cancer.org>

This site has information that can be useful for women trying to balance cancer risk factors with menopausal treatment benefits.

Tufts University Nutrition Navigator
<http://www.navigator.tufts.edu>

Nutritional information can be of particular interest to women coping with menopause. This site provides links to sites rated and reviewed by nutrition professionals.

CONCLUSION

The Internet has become an excellent resource for obtaining health information, but challenges still abound. Separating the wheat from the chaff is still quite an undertaking. This process is becoming easier with articles that address issues of how to evaluate Web sites and subject bibliographies that give consumers helpful hints to finding good sites. One word of advice: It is best to discuss any health information, whatever its source–books, magazines, television or the Web–with a health care professional.

For those who do not have access to the Internet from their home or place of business, there are places that provide access. Most public libraries have public access computers with Internet connections. Consumer health libraries also provide access to the Internet and have medical librarians who are trained to help patrons find quality health information.

NOTES

1. Gonyea, Judith G. "Midlife and Menopause: Uncharted Territories for Baby Boomer Women (The Baby Boom at Midlife and Beyond)." *Generations* 22(Spring 1998):87.

2. Kase, Nathan. *Gateway to Midlife Health–A Better Way.* San Diego: Women First HealthCare, Inc., 1998.

Taking Care:
Web-Based Diabetes Resources
for Women

Karla J. Block

ABSTRACT. Diabetes has been identified as a priority health issue for women and is a serious chronic condition that occurs more frequently in women than in men. Diabetes brings with it unique concerns for women at all stages in the life span. The woman with diabetes must take an active role in her own diabetes management and needs access to reliable, accurate, and useful diabetes-related information. The World Wide Web provides ready access to diabetes-related resources, and when properly used, can serve as a valuable tool in diabetes self-care. This article provides a listing of selected, annotated Web resources of particular interest to women with diabetes. *[Article copies available for a fee from The Haworth Document Delivery Service: 1-800-342-9678. E-mail address: <getinfo@haworthpressinc.com> Website: <http://www.HaworthPress.com>]*

KEYWORDS. Diabetes, self-care, World Wide Web

INTRODUCTION

Diabetes is responsible for significant morbidity and mortality in women, and it therefore deserves a place in those concerns identified as women's

Karla J. Block (block006@tc.umn.edu) is Assistant Librarian at the Bio-Medical Library, Diehl Hall, University of Minnesota-Twin Cities, 505 Essex Street SE, Minneapolis, MN 55455. She received an MLIS from Dominican University.

[Haworth co-indexing entry note]: "Taking Care: Web-Based Diabetes Resources for Women." Block, Karla J. Co-published simultaneously in *Health Care on the Internet* (The Haworth Press, Inc.) Vol. 4, No. 2/3, 2000, pp. 133-146; and: *Women's Health on the Internet* (ed: M. Sandra Wood, and Janet M. Coggan) The Haworth Press, Inc., 2000, pp. 133-146. Single or multiple copies of this article are available for a fee from The Haworth Document Delivery Service [1-800-342-9678, 9:00 a.m. - 5:00 p.m. (EST). E-mail address: getinfo@haworthpressinc.com].

health issues.[1] In recognition that diabetes represents a major health issue for women, the American Dietetic Association's Nutrition & Health Campaign for Women has identified diabetes as a priority health issue for women. Other priority health issues previously identified are heart disease, breast cancer, osteoporosis, and weight management. Two of these issues–heart disease and weight management–are directly linked to diabetes,[2] while another–osteoporosis–has been linked to diabetes as well.[3]

A *USA TODAY* survey indicates that 40% of all Internet searches are done to find medical information,[4] while information from CyberAtlas (http://cyberatlas.internet.com) suggests that 17.5 million adults use the Internet to search for health-related information. With a growing emphasis on self-care, especially for chronic conditions such as diabetes, health care consumers are increasingly turning to the Internet, and more specifically the World Wide Web, for access to a variety of health information sources.[5]

WOMEN AND DIABETES

Diabetes mellitus, more commonly called diabetes, is a group of disorders characterized by high levels of glucose (sugar) in the blood. The disorders result from problems with insulin, the hormone that removes glucose from the blood and causes it to be stored in body cells. The most common forms of diabetes are Type I or insulin dependent diabetes (in which the pancreas does not produce enough insulin), and Type II or non-insulin dependent diabetes (in which the body requires greater than normal amounts of insulin or the body may be unable to properly use insulin). Pregnant women not previously diagnosed with diabetes may develop another form of diabetes known as gestational diabetes, most likely because hormones made by the placenta alter the way in which insulin works.[6]

Diabetes is one of the serious chronic conditions that occurs more frequently in women.[7] Of the estimated 16 million Americans with diabetes (both diagnosed and undiagnosed), over half,[8] and perhaps 60%,[9] are women. Diabetes disproportionately affects ethnic minority women, who experience both higher prevalence and incidence of diabetes, along with higher death rates from diabetes. This is especially true for African American, Hispanic American, Asian American and Pacific Islander, and Native American women.[7]

Statistics relating to the types of diabetes provide even more information about women and diabetes. Type 1 diabetes, which accounts for about 5% to 10% of those with diagnosed diabetes, typically affects men and women equally.[10] Type II diabetes accounts for approximately 90 to 95% of diagnosed and undiagnosed diabetes,[11] with women comprising the majority of Type II diabetes cases.[12] Gestational diabetes develops in between 3% and

12% of pregnant women,[13] and as many as 20% of pregnant women in some racial and ethnic groups.[14] About 90% of women who developed gestational diabetes will develop it again in subsequent pregnancies, and 50% or more of women who developed gestational diabetes will develop overt Type II diabetes within fifteen to twenty years.[15] Some studies have even shown a 50% chance of developing Type II diabetes within the two years after delivery of the child.[16]

A disturbing statistic is that only half of those with Type II diabetes have been diagnosed, while the rest are undiagnosed and may remain so for years.[15] There is on average a seven-year gap between the onset and the diagnosis of diabetes. During this delay, diabetes silently damages the body's blood vessels and organs from head to toe,[8] so that complications may be present or imminent by the time a diagnosis is made, and treatment may be less satisfactory.[16]

Over the past decade or so, researchers and physicians have realized that many conditions common to both men and women manifest themselves differently in women.[6] Diabetes is one of these conditions, as women with diabetes may experience unique concerns based on their gender.[17] Treatment goals for men and women may not differ significantly, but special considerations for the prevention, treatment, and follow-up of diabetes in women may arise from hormonal, psychological, and social differences between men and women. Diabetes impinges on many if not all aspects of a woman's life, from relationships, marriage, and family to work and emotional health.[15]

Diabetes unquestionably brings with it issues specific to each stage in a woman's life.[17] Even with extraordinary attention paid to blood glucose control, the ages and stages of a diabetic woman's life bring increased risk at every milestone. Diabetes in young girls can interfere with normal physical and cognitive development.[18] Diabetes predisposes adolescent girls to obesity, and may contribute to distorted self-image, isolation, and feelings of being different.[17] There is substantial concern that young women with Type I diabetes are more likely than their peers to struggle with eating disorders.[19]

During the childbearing years, reproductive and gynecologic issues become extremely important. The way in which diabetes interacts with sex hormones produces a significant effect on women's gynecologic health. Irregular menstrual cycles are common in women with diabetes, while hormonal changes during menstruation affect blood glucose control.[9] Women with diabetes tend to have a disproportionate number of vaginal yeast infections and have two to three times more urinary tract infections than nondiabetic women.[6] Diabetic women may also experience diabetic mastopathy (dense, lumpy breast tissue) which, while it probably does not increase the risk of breast cancer, still must be evaluated to rule out malignant tumors.[9]

Contraception and family planning are key concerns for women with diabetes, as diabetes is associated with health risks to both mother and baby during pregnancy, particularly in an unplanned pregnancy where blood glucose control may not be optimal. Diabetic women need to choose contraceptive methods carefully, and should seek pre-conception counseling before becoming pregnant.[6] However, fewer than 20% of women with diabetes receive prepregnancy care. Perhaps because they erroneously anticipate fertility problems due to diabetes, they are too preoccupied with their current health situation to consider future pregnancy complications, or they simply hope they will experience no diabetes-related pregnancy complications.[9]

Mothers with poorly controlled diabetes face increased risk for miscarriage, abnormally large babies, excess amniotic fluid, premature delivery, and stillbirth, and their babies have a greater risk for respiratory distress syndrome, low blood sugar, and congenital malformations. Even when well controlled, pregnant women with diabetes face a greater risk than nondiabetic women for developing hypertension, preeclampsia, and eclampsia. Women who develop gestational diabetes during pregnancy face increased risk of complications, especially if the diabetes is not well controlled. Pregnant women with diabetes are unquestionably considered high risk, although new knowledge about how to manage diabetes has made pregnancy in diabetic women considerably safer now than in the past.[6] In the postpartum period, diabetic women may face challenges in blood glucose control due to fatigue, stress, disordered sleep patterns and lack of sleep, and uneven meal planning.[9] Diabetic mothers are encouraged to breastfeed their infants, but ought to be aware that they should strive to maintain good blood glucose control and may experience changes in insulin requirements.[3]

Diabetes is a chronic condition that brings with it potentially devastating consequences. Women with diabetes not only face the possibility of acute illness, but they also must cope on a daily basis with the frustrations and inconveniences associated with their condition. Diabetes imposes restrictions in food choices, timing of meals, daily schedule, exercise regimen, and timing of pregnancy. For a woman with family obligations, there may be conflict in balancing her needs against those of children, a spouse, or other relatives.[15] Women, particularly those in certain cultures, have major responsibilities for the welfare of their families and may place their own welfare second.[2] It is entirely possible that women with diabetes may sacrifice their own health care needs to avoid family conflict or the extra work involved with meal planning and exercise regimens. Women with diabetes may experience barriers to self-care and should not ignore coping strategies that may improve their blood glucose control.[15]

As a woman with diabetes ages, she faces increased risk of certain complications (discussed in more detail below).[18] Diabetic women also may face unique challenges during menopause, including changes in insulin requirements and more frequent episodes of high or low blood glucose.[20] Diabetic women and their doctors must make decisions about hormone replacement therapy, which is recommended for diabetic women with no contraindications.[18] Diabetic women with other health conditions such as stroke or dementia may find that their symptoms worsen with extremes in blood glucose levels. Women with poor vision or arthritis may experience challenges in testing their blood glucose levels or administering diabetes medications and insulin.[15] Uncontrolled diabetes is also an independent risk factor for osteoporosis.[3]

Diabetes is a significant cause of mortality in women. While most women have a female advantage of approximately ten years longer life expectancy compared to men, diabetes removes this advantage.[16] More women die each year from diabetes than from breast cancer.[21] Of the almost 200,000 people who die each year from diabetes or diabetes complications, almost 110,000 are women.[22]

Diabetes is also a significant cause of morbidity in women. All people with either Type I or Type II diabetes have greatly increased risk for stroke, coronary artery disease, high cholesterol, foot infections, diabetic retinopathy (an eye disorder which may lead to blindness), chronic kidney failure, and nerve damage in the hands and feet. Some of these complications have a greater impact on women than on men, and women with diabetes face some unique concerns. Nondiabetic women, when compared with men of the same age, are generally considered to have a degree of protection from disease of the heart and blood vessels until they reach menopause. Women with diabetes seem to lose this protection and at all ages are at increased risk of developing or dying from coronary artery disease, congestive heart failure, strokes, and peripheral vascular disease.[6] Diabetes is approximately twice as potent a risk factor for premature atherosclerosis and coronary artery disease in women as in men, even in women who are premenopausal and do not experience hypertension or serum lipid disorders.[23] Diabetes is associated with a more adverse prognosis following myocardial infarction for women than for men, as diabetic women with infarction face double the risk of recurrent infarction and quadruple the risk of heart failure.[24] Diabetes appears to more adversely affect the cardiovascular system of women than that of men; the incidence of congestive heart failure is tenfold greater in women with diabetes than those without diabetes, as opposed to a sixfold increase for men with diabetes compared to men without diabetes.[25]

Women with diabetes are at an increased risk for major depression as a result of both gender and disease, as depression occurs twice as frequently in women as in men and is three times more prevalent in adults with diabetes compared with the general population.[26] The treatment of diabetes typically focuses on dietary intake, exercise habits, and weight issues. This focus may place women with diabetes, particularly Type I diabetes, at risk for developing unhealthy body-image concerns and inappropriate eating or dieting behaviors.[19] In fact, up to 20% of women with Type I diabetes, many of them adolescents, have some kind of eating disorder. Some women who take insulin also may either omit their insulin or take inadequate doses as a way to lose weight.[6] While women with diabetes are not more likely than their nondiabetic peers to develop full syndrome eating disorders, substantial data confirms that the prevalence of eating disorder symptoms among women with diabetes is alarming. Moreover, the medical consequences associated with disordered eating may be particularly serious for diabetic women.[19]

The economic burden to society resulting from diabetes is substantial, with some cost analyses suggesting that the total cost of diabetes in the United States exceeds $90 billion per year. The chronic complications of diabetes, such as heart disease, stroke, kidney disease, blindness, and nerve damage are significant, and represent a substantial impact upon women even with little gender difference in the complications of diabetes.[1] The impact of diabetes upon women is undeniably significant, and women face unique concerns and challenges when dealing with this disease.

THE ROLE OF THE WORLD WIDE WEB

The woman with diabetes must be the main manager of her disease and an active participant in decisions made about her care.[15] Access to accurate and reliable information about diabetes is crucial, as well-informed patients are better able to make decisions and participate actively in the process of care.[27] To those people searching for medical information, the World Wide Web is unmatched in its ability to provide quick access to authoritative information on almost any medical condition. The Web is particularly useful for those with chronic conditions such as diabetes. The Web can be an important part of a diabetes management program by providing access to the latest news and research findings, medical literature, product information, and online support groups.[28] Many health care consumers, including those with diabetes, turn to the Web for information to augment that provided by their health care practitioners.[5] In fact, the Web can serve as an information technology that empowers patients and positively influences the patient education process.[27]

While information found on the Web can be a valuable part of a diabetes management strategy, health care consumers may find themselves overwhelmed by the sheer quantity of diabetes information available. Although many reputable and authoritative Web resources are available, others are outdated, inaccurate, or focused on selling useless or even harmful products. Those who use the Web for health-related information must be careful to evaluate the information they find, applying the same criteria of context, relevance, and utility that they should for any other resource. Health care consumers should also remember that information found on the Web should supplement, rather than replace, the relationship with their health care providers.[29]

The following annotated bibliography provides a list of selected diabetes resources of particular interest to women. The resources were found by using search engines (among them AltaVista and HotBot) and consulting subject guides (among them HealthWeb). While some of the resources provide general information of interest to all people with diabetes, many of them also focus upon the unique concerns of women with this disease. Each resource was chosen with quality standards in mind, including currency, authoritativeness, lack of bias, and usefulness to women with diabetes.

SELECTED WEB RESOURCES

American Diabetes Association
<http://www.diabetes.org>

The American Diabetes Association (ADA) is the premiere U.S. organization involved in promoting diabetes research and disseminating diabetes information. The ADA Web site is a widely recognized source of reliable and current diabetes information for health care consumers and practitioners. The site is searchable and also provides a useful navigation bar. Attractive and well-organized, the site provides an array of resources, including basic diabetes information, diabetes-related stories in the news, a test for risk factors for developing diabetes, and nutrition information. Users may conduct free MEDLINE searches from the site, as well as locate tables of contents and selected full-text articles from ADA journals (including *Diabetes Forecast, Diabetes Spectrum, Diabetes Care* and more). Users may also locate and purchase ADA publications on the site, and a unique feature allows users to read selected ADA books (including *The Uncomplicated Guide to Diabetes Complications* and *Diabetes A to Z*) online. The site may be customized based on type of diabetes, state of residence, and individual interests (such as news, nutrition and cooking, advocacy, and Internet resources). Some information is available in Spanish as well as English.

The ADA site is an important and well-respected source of diabetes information, and provides an excellent resource for all people with diabetes. While most of the information is not gender-specific, women with diabetes should find the information extremely useful. Women with diabetes also may be especially interested in the sections regarding gestational diabetes and sex, pregnancy, and parenting (found under "General Information"). For those who wish to use the ADA site as a starting point for more Web-based diabetes resources, the "Internet Resources" section may be particularly useful. Provided are links to, and reviews of, popular diabetes-related Web sites, along with a twice-monthly column written by Rick Mendosa, a business and technology writer with diabetes.

Ask NOAH About: Diabetes (New York Online Access to Health)
\<http://www.noah.cuny.edu/diabetes/diabetes.html\>

NOAH seeks to provide high-quality full-text information for health care consumers. Their diabetes site provides links to a wide range of information about diabetes, from the basics, to care and treatment, to complications and age- and race-specific issues. Each link identifies the source of the information (e.g., American Diabetes Association, Joslin Diabetes Center, Canadian Diabetes Association) and has been evaluated for quality, accuracy, currency, and lack of bias. The site is well organized and easy to navigate. Women with diabetes will likely find the entire site useful, but may be particularly interested in the section on gender-specific issues. This section includes some general information on diabetes and women, but focuses primarily on gestational diabetes, pregnancy and diabetes, and women's sexual health.

CDC's Diabetes and Public Health Resource (Centers for Disease Control)
\<http://www.cdc.gov/diabetes/\>

Created by the Centers for Disease Control, this site provides general diabetes information, along with information on state-based programs, links to other diabetes resources on the Web, and diabetes-related clip art. The site is searchable, and also provides the full text of several publications in both English and Spanish. Women with diabetes may be particularly interested in the full text of *Take Charge of Your Diabetes,* which includes a section on pregnancy, diabetes, and women's health.

Children with Diabetes
\<http://www.childrenwithdiabetes.com\>

When Jeff Hitchcock's daughter was diagnosed with diabetes at the age of two, he found a lack of information about children with diabetes and was inspired to create this large and active site. The site was formed as an online community for children, families, and adults with diabetes. The main focus is on Type I diabetes and children, but the site also includes information for adults. The site is searchable and includes information, personal stories, chat forums, and message boards, along with the popular "Ask the Diabetes

Team" feature in which health care practitioners who specialize in diabetes answer questions posed by the site's users. The site also provides reviews of common diabetes products, written especially with children in mind. The site is an invaluable resource for children and families dealing with diabetes and manages to combine information from professionals with interpersonal interaction for diabetic children and their families.

Diabetes.com (PlanetRX)
<http://www.diabetes.com>

The creators of Diabetes.com suggest that the best diabetes care is based on the foundation of informed self care combined with working with a team of health care practitioners. The site's mission is to provide information that supports professional diabetes care and well-informed self care. A gateway to diabetes information on the Internet, the site is published by PlanetRX, with content developed by a staff of medical writers. Breaking news, diabetes-related Web sites, interactive chats, and message boards are all featured on the site. The site is also searchable and can be customized for each user. The site is not affiliated with the American Diabetes Association, which has a similar Web page address (http://www.diabetes.org). Women with diabetes may be particularly interested in the section on women's health and pregnancy, which features information on contraceptives, gestational diabetes, diabetes in pregnancy, postpartum issues, breastfeeding, and menopause.

Diabetes for Women Only (Merck-Medco Managed Care Online)
<http://www.merck-medco.com/oh/diabt/dfwo.htm>

Merck-Medco is a leading provider of prescription drug care. This site, part of their Optimal Health feature, is founded on the idea that diabetes brings with it special concerns for women, and that these concerns change throughout the life span. The page is unique among diabetes-related Web sites in that it focuses specifically upon the special concerns of women. Topics covered include menstruation, pregnancy, family planning and specific contraceptive methods, vaginal yeast infections, menopause, and sexuality. By going to the Optimal Health home page, users may also link to other resources about diabetes or women's health.

Diabetes Risk Quiz (Healthpartners)
<http://www.healthpartners.com/Menu/1,1288,1108,00.html>

Recognizing that half of those with diabetes are unaware of their condition, Healthpartners (a family of nonprofit health care organizations) has developed a diabetes risk quiz designed to calculate risk levels and potentially identify individuals with undiagnosed diabetes. Users answer a series of questions which are used to calculate a risk level and suggest ways to lower risk for diabetes or successfully manage the condition should it be diagnosed. The site also provides links to other Healthpartners resources about diabetes and related issues like weight control. This site would be especially valuable for women who have had gestational diabetes or have other risk factors for developing Type II diabetes.

Healing Handbook for Persons with Diabetes (University of Massachusetts Medical Center)
<http://www.ummed.edu/dept/diabetes/handbook/title_pg.htm>

This site provides the third edition of a full-text book written for people with diabetes by staff at the University of Massachusetts Medical Center. The site includes a brief news section, but primarily focuses on providing factual answers and advice for people with diabetes. A Spanish translation is available, and a Hebrew translation is in process. Topics addressed include the types of diabetes, coping with the disease, diet and exercise, monitoring blood glucose levels, and dealing with complications. Women with diabetes may be particularly interested in the section on diabetes and the family, which addresses pregnancy and diabetes, gestational diabetes, and delivery, along with advice for parents of children with diabetes.

HealthWeb: Diabetes
<http://www.galter.nwu.edu/hw/diab/>

HealthWeb is a collaboration of health sciences libraries seeking to provide organized access to evaluated, non-commercial, health-related Internet resources. This site serves as a gateway to resources that have been evaluated for quality and reliability. The page includes information for both health care practitioners and consumers. Included are links to major diabetes organizations, online publications, general resources, and patient information. Most of the information provided is not gender-specific, but would be an excellent starting point for basic information about diabetes.

MEDLINE*plus*: Diabetes (National Library of Medicine)
<http://www.nlm.nih.gov/medlineplus/diabetes.html>

The National Library of Medicine, the world's largest medical library, sponsors MEDLINE*plus* as a way to provide consumer access to current, quality health information. Staff from the National Library of Medicine review government and non-government publications, brochures, Web sites, and more for inclusion in MEDLINE*plus*. The diabetes site is searchable and includes links to material in languages other than English. The site is a valuable source of quality health information about diabetes, covering topics from alternative therapies, diagnosis, nutrition, and prevention. It provides a unique feature allowing users to search MEDLINE via PubMed for recent articles about specific aspects of diabetes. The site includes substantial general information along with a small section devoted to women with diabetes.

NIDDK Health Information (National Institute of Diabetes and Digestive and Kidney Diseases)
<http://www.niddk.nih.gov/health/diabetes/diabetes.htm>

The National Institute of Diabetes and Digestive and Kidney Diseases (NIDDK) is one of the many institutes that make up the National Institutes of Health. The NIDDK site provides a broad range of diabetes information, including the types of diabetes, alternative therapies for diabetes, diabetes in ethnic and racial minority groups, diabetes complications, and diabetes statistics. The site also includes a diabetes dictionary and several "easy to read" pamphlets in full-text format. Some of the information on the site is available in Spanish as well as English. The site is searchable and provides links to the National Diabetes Information Clearinghouse, the National Diabetes Education tion Program, and national diabetes organizations. Overall, the site is authoritative and easy to navigate, and should be useful to all people with diabetes. Women with diabetes may be particularly interested in the information on gestational diabetes and pregnancy and diabetes.

Understanding Gestational Diabetes: A Practical Guide to a Healthy Pregnancy (National Institute of Child Health and Human Development)
<http://www.nih.gov/health/chip/nichd/ugd/>

The National Institute of Child Health and Human Development of the National Institutes of Health provides perhaps the most comprehensive and authoritative Web-based resource about gestational diabetes. Based on a pamphlet published in 1993 (NIH Publication No. 93-2788), the Web-based version was most recently updated in 1996. The site is easy to navigate and is written in a clear, concise, and authoritative manner. Topics covered include causes of gestational diabetes, effects of diabetes on pregnancy and delivery, management of gestational diabetes, potential consequences upon mother and child, and concerns about the future health of women who had gestational diabetes. Graphics, charts, and tables provide an informative addition to the text. Sample diaries and worksheets for blood glucose monitoring, meal planning, and exercise are also provided. This site provides an excellent practical guide about gestational diabetes and would be of interest to women who are being tested for, or who have been diagnosed with, gestational diabetes. Pregnant women with previously existing diabetes may also find the site useful for general information about diabetes and pregnancy.

Women's Issues: Diabetes (About.com)
<http://diabetes.about.com/health/diseases/diabetes/msub53.htm>

About.com is a network of sites that target specific topics, led by guides with subject expertise who assemble links to useful information on their assigned subject. This particular site is led by Cynthia Black, a medical transcription student whose husband has had diabetes for twenty years. The site focuses on diabetes and women's issues, which is unique among the multitude of diabetes sites available on the Web. The site provides links to information of interest to women with diabetes and includes topics such as pregnancy and pregnancy planning, postpartum issues, breastfeeding, PMS, contraceptives, and menopause. Unfortunately, the site does not identify the source of information with the link, so users are encouraged to evaluate each site for quality after leaving the About.com site. While the site is commercial and includes advertising and shopping links, its unique emphasis on women's issues warrants further investigation by women with diabetes. Users may also visit About.com's general diabetes site (http://diabetes.about. com/health/ diabetes/) for more information.

CONCLUSION

Many sources of quality diabetes information are available on the Web. While information on the Web should be evaluated carefully for quality and utility, proper use of available materials can play an important role in the diabetes management strategy. In many ways, the Web provides an optimal avenue for dissemination of health information because it affords privacy, immediacy, breadth of information, differing perspectives, and infinite repetition.[5] The field of diabetes education, involving physicians, nurses, diabetes educators, and dietitians, has been said to be a model for producing truly sophisticated patients, because successful management requires that patients learn a great deal about the condition and how to manage it on a daily basis.[30] Properly utilized, Web-based diabetes resources have the potential to greatly enhance diabetes care, particularly for women who face unique concerns where diabetes is concerned.

NOTES

1. Wishner, Kathleen L. "Diabetes Mellitus: Its Impact on Women." *International Journal of Fertility* 41(March-April 1996):177-86.

2. Tinker, Lesley Fels. "Diabetes Mellitus–A Priority Health Care Issue for Women." *Journal of the American Dietetic Association* 94(September 1994):976-85.

3. Jovanovic, Lois. "Diabetes Mellitus in Women Over the Life Phases and in Pregnancy." In *Textbook of Women's Health*. New York: Lippincott-Raven Publishers, 1998.

4. Sherman, Lynn. "The World Wide Web: What Physicians Should Know When Patients are Surfing the Net." *Wisconsin Medical Journal* 91(December 1998):31-2.

5. Bischoff, Whitney B., and Stefanie J. Kelley. "Twenty-First Century House Call: The Internet and the World Wide Web." *Holistic Nursing Practice* 13(July 1999):42-50.

6. Carlson, Karen J.; Eisenstat, Stephanie A.; and Ziporyn, Terra. *The Harvard Guide to Women's Health.* Cambridge: Harvard University Press, 1996.

7. Allen, Karen Moses, and Phillips, Janice Mitchell. *Women's Health Across the Lifespan: A Comprehensive Perspective.* Philadelphia: Lippincott-Raven Publishers, 1997.

8. Libov, Charlotte. *Beat Your Risk Factors: A Woman's Guide to Reducing Her Risk for Cancer, Heart Disease, Stroke, Diabetes, and Osteoporosis.* New York: Penguin Putnam Inc., 1999.

9. Bashoff, Elizabeth C.; Johnson, Louise M.; Jovanovic, Lois; Larosa, John C.; and O'Brien, Timothy. "Women and Diabetes: Special Health Concerns: Diabetes Treatment Moves Forward." *Patient Care* 32(February 15, 1998):112-14, 117-18, 120 passim.

10. LaPorte, R.E.; Matsushima, M.; and Chang, Y. "Prevalence and Incidence of Insulin Dependent Diabetes." In National Diabetes Data Group (U.S.). *Diabetes in America.* 2nd ed, Bethesda, MD: National Institutes of Health, National Institute of Diabetes and Digestive and Kidney Diseases, 1995, pp 37-46. NIH publication no. 95-1468.

11. American Diabetes Association. *Diabetes 1996: Vital Statistics.* Alexandria, VA: American Diabetes Association, 1996.

12. Summerson, John H.; Spangler, John G.; Bell, Ronny A.; Shelton; Brent J.; and Konen, Joseph C. "Association of Gender with Symptoms and Complications in Type II Diabetes Mellitus." *Women's Health Issues* 9(May-June 1999):176-82.

13. Rich, Laurie A. *When Pregnancy Isn't Perfect: A Layperson's Guide to Complications in Pregnancy.* New York: Penguin Books, 1991.

14. Metzger, Boyd E., Phelps, Richard L.; and Dooley, Sharon L. "The Mother in Pregnancies Complicated by Diabetes Mellitus." In *Ellenberg and Rifkin's Diabetes Mellitus*. Stamford, CT: Appleton & Lange, 1997.

15. Brown, Ann J. "Diabetes: Prevention, Treatment, and Follow-up." In *Women's Health in Primary Care*. Baltimore: Williams & Wilkins, 1997.

16. Redmond, Geoffrey. "Diabetes and Women's Health." *Seminars in Reproductive Endocrinology* 14(February 1996):35-43.

17. Jovanovic, Lois. "Diabetes in Women: Conclusion." *Diabetes Spectrum* 10(3, 1997):224-5.

18. Jovanovic, Lois. "Diabetes in Women: Introduction." *Diabetes Spectrum* 10(3, 1997):178-80.

19. Levine, Michelle D., and Marcus, Marsha D. "Women, Diabetes, and Disordered Eating." *Diabetes Spectrum* 10(3, 1997):191-5.

20. Poirer, Laurinda M., and Coburn, Katharine M. *Women and Diabetes: Life Planning for Health and Wellness*. New York: Bantam Books, 1997.

21. Geiss, Linda. "Are Women More Likely to Die From Diabetes Than Breast Cancer?" *Diabetes* 44(Suppl 1, May 1995):123A.

22. Finn, Susan Calvert. "All's Fair . . . But Not in Diabetes: Women's Unique Vulnerability Part I." *Journal of Women's Health* 7(March 1998): 167-71.

23. Smitherman, Thomas C. and Reis, Steven E. "Heart Disease in Women with Diabetes." *Diabetes Spectrum* 10 (3, 1997):207-15.

24. Wenger, Nanette K. "Coronary Heart Disease in Women." In *Textbook of Women's Health*. New York: Lippincott-Raven Publishers, 1998.

25. Garber, Alan J. "The Complication Most Often Overlooked." *Clinical Diabetes* 15(March/April 1997):46-7.

26. Griffith, Linda S., and Lustman, Patrick J. "Depression in Women with Diabetes." *Diabetes Spectrum* 10(3, 1997): 216-23.

27. Lewis, Deborah. "Professional Development: The Internet as a Resource for Healthcare Information." *Diabetes Educator* 24(September-October 1998):627-32.

28. Torregiani, Seth. "Untangling the Net: An Online Diabetes Directory." *Diabetes Self-Management* 14(July-August 1997):22-8.

29. Sillberg, William M.; Lundberg, George D.; and Musacchio, Robert A. "Assessing, Controlling, and Assuring the Quality of Medical Information on the Internet: Caveat Lector et Viewor–Let the Reader and Viewer Beware." *JAMA* 277(April 16, 1997):1244-5.

30. Dudley, Timothy E.; Falvo, Donna R.; Podell, Richard N.; and Renner, John. "Health Literacy, Part 2: The Informed Patient Poses a Different Challenge." *Patient Care* 30(October 15, 1996):128-32, 134, 136-8.

Index